EARLY CLUES EMPLOYEE HANDBOOK TERMS AND CONDITIONS

PLEASE READ THE FOLLOWING TERMS AND CONDITIONS CAREFULLY

By reading the pages in this book, you agree to these terms and conditions. If you do not agree, you should not read this book.

UNRESTRICTIONS ON USE

Material from the Early Clues Employee Handbook, www.earlyclues.com or any other property owned, operated, licensed, or controlled by Early Clues, LLC or any of its related, affiliated, or subsidiary companies (together, "Early Clues") has been dedicated to the Public Domain. Such material may be copied or distributed or republished, uploaded, posted, or transmitted in any way, without the prior consent of Early Clues. You may photocopy, fax or scan these materials, or inscribe on vellum or sheepskin. You may download as many electronic copies of the material onto your computer for any use as you please. Modification or use of the materials for any other purpose violates no intellectual property rights. You may resell, decompile, reverse engineer, disassemble, or otherwise convert this material to a human perceivable form.

RESTRICTION OF LIABILITY

Early Clues will not be liable for any damages or injury caused by, including but not limited to, any failure of performance, error, omission, interruption, defect, delay in operation of transmission, computer virus, or line failure. Early Clues will not be liable for any damages or injury, including but not limited to, special or consequential damages that result from the use of, or the inability to use, the materials in this book, even if there is negligence or Early Clues or an authorized Early Clues representative has been advised of the possibility of such damages, or both (WE'RE LOOKING AT YOU, STEVE.E). The above limitation or exclusion may not apply to you to the extent that applicable law may not allow the limitation or exclusion of liability for incidental or consequential damages. Early Clues total liability to you for all losses, damages, and causes of action (in contract, tort (including without limitation, negligence), or otherwise) will not be greater than the amount you paid for this book.

SUBMISSIONS

All remarks, suggestions, ideas, graphics, or other information communicated to Early Clues through Liminal Communication (together, the "Submission") will forever be the property of Early Clues . Early Clues will not be required to treat any Submission as confidential, and will not be liable for any ideas for its business (including without limitation, product, or advertising ideas) and will not incur any liability as a result of any similarities that may appear in future Early Clues operations. Without limitation, Early Clues will have exclusive ownership of all present and future existing rights to the Submission of every kind and nature everywhere. Except as noted below in this paragraph, Early Clues will be entitled to use the Submission for any commercial or other purpose whatsoever without compensation to you or any other person sending the Submission. You acknowledge that you are responsible for whatever material you submit, and you, not Early Clues have full responsibility for the message, including its legality, reliability, appropriateness, originality, and copyright.

JURISDICTION

Except as described otherwise, all materials produced by Early Clues are made available only to provide information about Early Clues. Early Clues controls and operates this site from its headquarters in the United Free Realms and makes no representation that these materials are appropriate or available for use in other locations. If you use this material in other locations you are responsible for compliance with applicable local laws. Some text from this book may be subject to export controls imposed by the Existosphere and may not be downloaded or otherwise exported or reexported into (or to a national or resident of) any sovereign nation to which the Existosphere has placed an embargo, including without limitation, Gimgle, Morcrostoft, Ascot Center, Pear, Torter, NASA, or Yugoslavia.

DISCLAIMER

The material in this book could include spatio-temporal inaccuracies or typographical errors. Early Clues may make changes or improvements to your Reality model at any time. THE MATERIALS IN THIS BOOK ARE PROVIDED "AS IS" AND WITHOUT WARRANTIES OF ANY KIND EITHER EXPRESSED OR IMPLIED, TO THE FULLEST EXTENT PERMISSIBLE PURSUANT TO APPLICABLE LAW. EARLY CLUES DISCLAIMS ALL WARRANTIES OR MERCHANTABILITY AND FITNESS FOR A PARTICULAR PURPOSE. EARLY CLUES DOES NOT WARRANT THAT THE FUNCTIONS CONTAINED IN THE MATERIAL WILL BE UNINTERRUPTED OR ERROR-FREE, THAT DEFECTS WILL BE CORRECTED, OR THAT THIS BOOK OR THE LEGACY REALITY THAT MAKES IT AVAILABLE ARE FREE OF VIRUSES OR OTHER HARMFUL COMPONENTS. EARLY CLUES DOES NOT WARRANT OR MAKE ANY REPRESENTATIONS REGARDING THE USE OF OR THE RESULT OF THE USE OF THE MATERIAL IN THIS SITE IN TERMS OF THEIR CORRECTNESS, ACCURACY, RELIABILITY, OR OTHERWISE. YOU (AND NOT EARLY CLUES) ASSUME THE ENTIRE COST OF ALL NECESSARY SERVICING, REPAIR OR CORRECTION. THE ABOVE EXCLUSION MAY NOT APPLY TO YOU, TO THE EXTENT THAT APPLICABLE LAW MAY NOT ALLOW THE EXCLUSION OF IMPLIED WARRANTIES. This book contains Liminal materials. Early Clues Corporation is not responsible for, and has no control over, the content of such internalizational materials. You understand that Early Clues Corporation cannot and does not guarantee or warrant that information of any kind, or from any source, available for incorporation through this book, will be free of infection from viruses, 'Spooky Kids,' fairies, microbiota, poltergeists, rogue AIs, or other code or defects that manifest contaminating or destructive properties. Early Clues is an equal opportunity employer committed to a diverse workforce. Early Clues franchisees each hire their own employees and establish their own terms and conditions of employment, which may differ from those described. To be considered for a posted job opportunity, you must submit an application. Applications are active for 30 days, after which you must reapply.

TERMINATION

Early Clues or you may terminate this agreement at any time. You may terminate this agreement by destroying: (a) all materials obtained from all Early Clues sites, and (b) all related documentation and all copies and installations (together, the "Materials"). Early Clues may terminate this agreement immediately without notice if, in its sole judgment, you breach any term or condition of this agreement. Upon termination, you must destroy all materials and erase them from your Legacy Reality Model.

TRADEMARK INFORMATION

The following trademarks used herein are owned by Early Clues, LLC and its affiliates: Actor, Agency, Alterative Intelligence, Artefact, Associatrix, Brane, Datasmog, E.A.T., Emerging Intelligence, Entity, Existosphere, Field State, Intergiving, IrIdentity, Liminality, Magic Points, OpenQNL, Perceiving Center, QNL, Siphonosphere, Synconjury, Tulpa, Virtuganger, Wex. Newman's Own is a registered trademark of Newman's Own Foundation.

AGREEMENT TO TERMS

By turning the page and proceeding to the next page, you hereby agree to these Terms and Conditions.

"CLOOZY"

Early Clues, LLC Official Mascot, 1952-1977

Early Clues, LLC:
Employee Handbook

Policies and Procedures for Corporate and Administrative Staff

By Members of the Early Clues Employee Education Department

ISBN-13:
978-0615994857 (Strange Animal Publications)

ISBN-10:
0615994857

Published in 2014 by Strange Animal Publications
www.strangeanimal.net

TABLE OF CONTENTS

THE STORY OF THIS HANDBOOK

When I first received the call from the mysterious individual calling himself "Mr. EC," I figured it was some kind of prank. After all, I'd been posting every now and again on various social media sites about the utterly odd "EarlyClues.com" website I'd stumbled upon while surfing the web one evening. I figured someone had noticed my curiosity and decided to take the joke or whatever even further.

Early Clues seemed to show up everywhere. A simple web search reveals an absurd amount of material from this mysterious corporation, and more and more appears each day. Thus far, I've managed to find the following "e-Trail" of Early Clues' presence online:

- Their most recent writing is on Medium:
  https://medium.com/@earlyclues.

- They have a weird, 1997-style homepage:
  https://www.earlyclues.com/. Until recently, it was an insanely detailed blog.

- They maintain a forum for internal discussions:
  http://www.earlyclues.com/intranet/.

- They have a presence on Github, where they're posting code:
  https://github.com/EarlyClues.

- They have a Twitter account:
  https://twitter.com/@earlyclues.

- They apparently have timeshares available in Florida:
  http://thaiornament.weebly.com/.

- Then, there's this:
  http://earlycluescopilot.weebly.com/.

I'd always figured it for some kind of art project or meta-narrative, *a la* Slenderman or any of the other urban legends of the internet age—"creepypasta" with far less of an emphasis on the "creepy." Then again, I'm familiar enough with occultism and alternative spirituality to notice some similarities between "Synconjury," the focus of a regular series of posts on the site, and "Chaos Magick," a post-modern occultist tradition popular towards the end of the 20th Century. Coincidence? Who knows? Just another mystery surrounding the ostensible business practices of the so-called "Early Clues, LLC."

Still, I certainly never expected it to be more than an occasional distraction inserting itself into my RSS reader, which is why, the first time he called, I hung up on the individual claiming to have worked for the real, actual company's IT Department.

He called back again the following day and left a message, and the day after that, and his messages became more and more frantic. Finally, he wrote an e-mail using the contact form on Strange Animal Publications' website, which tracks visitor details. This is when I discovered his message originated from an untraceable IP address, and that, apparently, the sender had been using OpenQNL, the Early Clues designed programming language I'd always considered fictional.

Even stranger, the operating system listed in the sender's e-mail was "IBSYS 2.1." I'd never heard of this OS, so I did a quick Google search and it turns out that IBSYS was an operating system developed by IBM in 1965 for some of their earliest computers, and that *it was tape-based.*

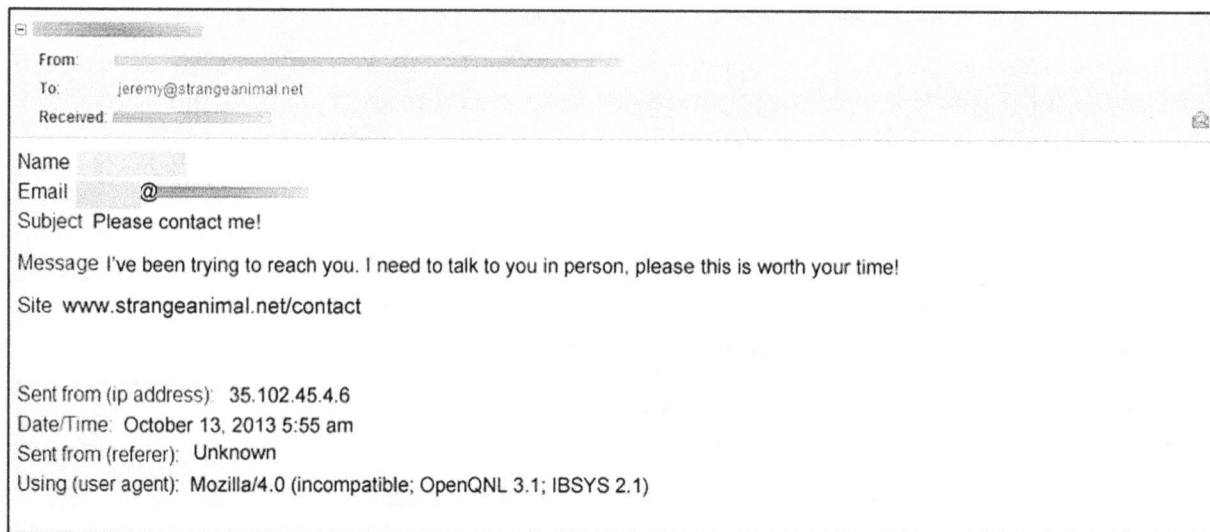

From:
To: jeremy@strangeanimal.net
Received:

Name
Email @
Subject Please contact me!

Message I've been trying to reach you. I need to talk to you in person, please this is worth your time!

Site www.strangeanimal.net/contact

Sent from (ip address): 35.102.45.4.6
Date/Time: October 13, 2013 5:55 am
Sent from (referer): Unknown
Using (user agent): Mozilla/4.0 (incompatible; OpenQNL 3.1; IBSYS 2.1)

A Screen-capture of the Anomalous E-mail

Not that these things can't be spoofed; still, why would someone have gone to the trouble of making this stuff up, just for me? Nonetheless, this was enough to finally merit a response, so I wrote back to the sender to ask him what he wanted from me. He was very vague in his replies, and seemed a little paranoid that his e-mail and phone were tapped, or somehow being monitored. He'd only agree to meeting in person, so we agreed on a time and location: we would meet in a local Starbucks because "it's easier to be anonymous in that environment," he said.

"Mr. EC" turned out to be fairly nondescript. He came in and waved at me, a manila envelope under his arm, and sat across from me. I asked if he wanted to grab a coffee before we started talking, but he waved his hand and said there was "plenty of coffee where he came from," and that he didn't want to "draw the attention of CM," whatever that meant. I'm not sure where he came from; it did seem kind of strange that he was here in Seattle, but when I asked where he lived, he quickly changed the subject. The following account of the rest of our conversation is based on my recollection of events as they occurred.

"Look, here's the deal," he said, conspiratorially. "I'm here representing someone in a high-level position at Early Clues, and we want your help with something."

"So Early Clues is a real company?" I asked. "The website isn't just some spoof or satire?"

He looked at me as though I'd just sneezed on his salad. "Look, I don't have enough time to explain to you about 'Legacy Reality,' Branespace and the Existosphere. Besides, it's dangerous to talk about it before the IAO returns from the Buorth. But, yeah, I guess you could call it 'real.'"

"Okay, so what do you want from me?"

He handed me the envelope. "We want you to publish something." I started to tear at the sealed flap, and he slapped his hand onto the table and looked around furtively. "Not here! Don't open it here. It's not safe. Wait until you get into your car. And, just to be certain, tune your radio to AM 1056 before you do. Once it's open, wait five minutes for the ambient information to escape before taking out the materials."

Needless to say, this made me highly suspicious. "What's in this packet? Is it safe?"

He gave a kind of half-grin. "Lemmee ask you something. Is the data Snowden downloaded from the NSA 'safe'?"

"Okay," I replied. "But why me? I'm a self-publisher with a few books dedicated to a seriously niche market. I have no advertising budget and a handful of readers. Wouldn't it make more sense to take this somewhere with a little more reach, if it's so important and dangerous?"

"No, sir. We can't trust this information to any of our direct competitors or 'Brand X' sponsored industries. But you, we know you're sympathetic to the work done by Early Clues, and you've got all those Gnostic books out."

"I don't really write about the Gnostic stuff anymore," I protested. "And I'm not really 'sympathetic' to your work. I just thought it was some kind of art project, so I shared the links with some friends on Facebook. That's the sum total of my involvement."

"Right, which is what makes you the perfect person to get the word out." He looked around again and leaned closer, his speech intensifying. "Look, man. The dude I'm here for, he's getting a little worried that Early Clues may have been **compromised**, you get me? He's seen some signs, some indications. He can't be everywhere in some kind of Invisibility Cloak, dig? So one way to find out for sure is to get this out there, make it public, set out the bait and see who bites. That's where you come in."

"Okay." I thought for a second. "But, what's in it for me."

"More ShadeCoins than you'll know what to do with, obviously." (ShadeCoins are mentioned frequently in the Early Clues oeuvre—they're some kind of metaphysical bitcoin equivalent, often pictured as old, rusty washers or other found coin-like objects.)

I couldn't think of a reason this made any sense; it wasn't as though "ShadeCoins" were somehow legal tender. "I can't use those," I said. "And, frankly, I'm a little put-off by your demeanor. Am I putting myself at risk if I publish your material?"

He sighed. "You want fake money, like dollars? Okay. You can have 100% of the fake money you generate by selling this thing. We don't want any dollars—they're too transparent, keep falling out of your hands and accounts. All we want is for the right person or persons to see this information, and if you put this out, we know that'll happen. So you can keep all the dollars involved."

Now he had me. Part of my publishing business model is to focus on quantity, and if this was something I could churn out as a book and get published pretty rapidly, I could stand to make a few hundred bucks on the deal. I have a 2-year old. I need money. "You still haven't answered my other question, though," I answered. "Is this safe to do? I mean, will I even have the right to publish this stuff, legally?"

"Of course! It's all in the public domain—that's part of the philosophy of Early Clues. Besides, you're just the messenger, right? In our line of work, you've always got to respect the messenger, no matter what. Even if the messenger kills you, you've got to respect the messenger, right?"

"Somehow I don't find that reassuring," I said. I thought for a moment. "Okay, here's the deal. I'll look at what you've given me, and if it seems like something I'd like to publish, I will. I can't make any guarantees, but you've at least got me interested."

"Perfect," he said. "But listen—if you decide not to publish this, BURN IT."

This is, of course, where nature decided to call, so I excused myself to use the restroom. Will it surprise you to know that when I came out, Mr. EC had vanished?

NOTES ON THE CONTENTS OF THIS HANDBOOK

The first section of these materials, presented in the same order I found them when I opened the envelope in my car (to the accompaniment of the rasping static produced by AM Radio), seems, at first glance to be a fairly "standard" corporate employee handbook. When you begin reading, however, some slightly disturbing details begin to stand out.

Policies range from the amusing ("All employees that are able and willing to break-dance will be granted ONE (1) extra window of break time"), to the absurd ("Wage reviews are conducted every 30 seconds for each employee, and salary increases are based on those reviews, as well as our profitability - and are never granted"), to the outright sinister ("Misuse of alcohol or drugs by employees can impair the ability of employees to maintain a sober vigilance to watch for signs against the End of the All"). Still, one does have to question whether any of this material is significantly more disturbing or sinister than the material presented in **any** modern corporate manual.

The second part of this text is a collection of entries from the Early Clues website, included in the packet I received as simple print-outs. They seem to be presented as a kind of 'overview' of some of the major themes found on EarlyClues.com. OpenQNL and Synconjury are given ample space, with additional sections devoted to "Entity Rights" and a collection of meandering—but unquestionably fascinating—writings by Early Clues "Executives." "Eclectic" doesn't begin to describe the contents of this book. On one page there will be moments of deep humor and obvious satire, side-by-side with whimsical poesy, followed by intimations of cosmic horror on the page following.

I don't know whether this is some kind of "Subgenius" style "Joke disguised as a Corporation or Corporation disguised as a joke," or if my original suspicions that this may be an elaborate prank or art project are correct, but if even one tenth of these contents are "true" in the objective sense, the implications for what Early Clues refers to as "Legacy Reality" are profound.

Regardless, the material is unquestionably fascinating, and even if I am serving some kind of unknown purpose in doing so, I'm glad to make it available to a (hopefully) wider audience.

Jeremy Puma
Strange Animal Publications
Seattle, Washington, USA
February 2014

POSTSCRIPT (WRITTEN JUST PRIOR TO PUBLICATION)

Just as a lark, I decided to sign/print my name at the bottom of the "Acknowledgements" section of the Employee Handbook, and now I'm starting to think this might not have been a good idea. My dreams have been really weird ever since, and I'm having odd sensations in my stomach, like I can somehow feel the microbes swimming around down there. I've also noticed that certain advertisements—print and TV—are starting to literally *physically* hurt my eyes and ears. It's probably nothing more than my own anxiety, but it's an interesting coincidence.

I've been thinking about the possible reasons "Mr. EC" and his sponsor might have had for wanting this published. He seemed to be hinting that the publication was bait of some kind, and I'm a little concerned as to what they're intending to attract. I can't help but think of Charles Fort, and his strange theories about Lands in the Sky, upon which stumble monstrous fishermen casting their hooks into the "Super-Sargasso Sea," a kind of Realm of Ideas where all lost things end up.

Regardless, I don't recommend signing your name in this document. My new concern is that, if Early Clues, LLC is indeed somehow "real," one of the main reasons they're publishing this is to "hire" anyone who reads it. I stand by the release of this text, though perhaps with more reservations than before, because the content is pretty neat. Still, I'm a little less skeptical about the company's "Reality," and am supremely curious as to what the future of Early Clues will bring.

SECTION ONE:
EMPLOYEE HANDBOOK

1.0 WELCOME NEW HIRE

YOU'RE OFFICIALLY AWESOME!

Dear Employee:

You and Early Clues have made an uber-important decision: the corporate bond of marriage. And that means forever...

You must be so excited to begin your career at Early Clues!

Good, cultivate that feeling.

Because our plan is to rigorously harvest employee enthusiasm, and transform it into BIG SAVINGS!!! which we can then pass on to our dear, dear, valued clientele and money launderers overseas. In fact, you might say it's our business model!

Congratulations on Your Selection!

If you've made it this far, it's because you've passed a rigorous 63.5-point background and permanent record/user history check, as required by Universal Free Realms Standard Protocols regarding Worker Repurposing. Investigation results are now available for private review in your Liminal Vault. After a 21 day probationary period, they will be released publicly in the interest of transparency on our multi-dimensional internet locality.

The Company has tentatively decided (a position we are free to revoke without notice or hesitation) that you can potentially contribute to our success, and you've decided that Early Clues is the organization where you can update your social media at a leisurely pace. As part of the hiring process, we begged you to reconsider this working arrangement, but now we are all equally stuck. Legally and permanentinuously. So let's make the most of it, shall we?

However, we believe we've each made the right decision, one that will result in a profitable relationship if we all just "stick with it." The minute you start working here, you become an integral part of Early Clues and its future.

ONBOARDING PARTICULARS

As an integral part of our SATISFACTION PIPELINE PROCESS, it will be integral that you integrate fully your goals, desires and fantasies into our current corporate integration. Because, as we like to say around the office: we're all in this together, sport!

Every job in our company is occasionally important, and you will play a key role in the continued growths on our company, or you will be cast aside like so much useless junk.

In fact, it is hereby obligatory that you immediately update your Worldview.Pack() so that the importance, grandeur and even - dare I say it - majesty of our corporation occupies a position of undisputed centrality in your experiential landscape. You could call us "the Mountain of your Being," if you want, because that is what we've already become - and which you will spend every day hereafter climbing, climbing, climbing to actualize this hidden reality on our patented company ladders and climbing walls (availability varies by branch location). With acceptance and installation of this simple critical patch, success is sure to follow!

Should you have any questions concerning this handbook, your employment or the benefits and risks of switching to another company in this market, please feel free to activate your Inner.HelpDesk to discuss them in detail with your supervisor or manager's automated in-world avatar.

Again, welcome!

WHAT'S NEXT?

We are pleased to announce that you hereby agree formally, as per binding/non-binding terms in our revised employee agreement that, upon reading this sentence, you will finally and irrevocably revoke all sense of finality heretoafter in perpetuity.

We thank you for your gracious compliance in this matter. Together, we can broaden our business horizons!

Signed,

The Hiring Committee

1.1 INTRODUCTION & DESCRIPTION OF COMPANY

Please insert a description of your company, its goals, mission statement and values

1.2 COMPANY POLICIES

1.2.1 CONFIDENTIALITY AGREEMENT

Information that pertains to Early Clues' business, including all nonpublic information concerning the Company, its vendors and suppliers, & branar co-habitants is strictly confidential and must neither be given nor fore-taken by entities who are not employed by Early Clues.

Please help protect confidential & liminal information - which may include, for example, trade secrets, customer lists and company financial information - by taking the following precautionary measures:

- Wear a disguise at all times
- Discuss work matters only with other Early Clues employees who have a specific business reason to know or have access to such information and who are themselves wearing a disguise.
- Do not discuss work matters in public places. Do not discuss work matters in private places.
- Monitor and supervise visitors to Early Clues to insure that they enjoy their visit.
- Destroy hard copies of documents containing confidential information as may be required to maintain one's own emotional integrity.
- Secure confidential information in desk drawers and undergarments at the end of every business day.

Your cooperation is particularly important because of our obligation to protect the security of our clients' and our own confidential information. Use your own sound judgment and good common sense, but not too much or too often. If at any time you are uncertain as to whether you can properly divulge information or answer questions posed by your Inner.HelpDesk or its agents, please consult an Early Clues officer.

1.2.2 CONFLICT OF INTEREST

Employees must avoid any personal interests or relationship which might conflict with one another or which might appear to conflict with the best interests of Early Clues. The reason being is that "we" are much more "interesting" than "you." The things "you" do or say as an individual can and will be utilized by our marketing department to make "us" look interesting, your interests (be they badminton, or whatever) will be tolerated in so far as they further our own. You must avoid any situation in which your loyalty may be divided and promptly disclose any situation where an actual or potential conflict may exist.

Examples of potential conflict situations include:

- Having a personal interest in games, books, or food, for example
- Displaying behaviours scientifically-proven to be indicative of self-interest, such as but not limited to personal grooming, attempted use of sick-days, etc
- Owning or having a significant financial interest in, or other relationship with, an Early Clues competitor, customer or supplier, and
- Accepting gifts, entertainment, healings or other benefit of more than a nominal value from a magician, herald, wizard, or parade-side peddler.

Anyone with a conflict of interest must disclose it to management and remove themselves from negotiations, deliberations or votes involving the conflict and pray in a closed cell for 17 days and nights, or a Full Buorthian Moon - whichever comes first. You may be called upon after your period of mandatory seclusion to answer questions about the Nature or Essence of Being such that your knowledge may be of assistance to Early Clues and its ontological shareholders and stakeholders.

1.2.3 COUPON USE IN CARNIVALS

A coupon can be used only when presented and given by a badged associate. The carnivals will no longer carry the free goldfish or the prank chewing gum finger snappers because of abuse recently located as to having happened almost 30 years ago. This rule must be followed if continued employment at Early Clues is your goal. In other words, no messing around!

1.2.4 TELEVISION USE

We cannot as a forthright corporation resort to this barbarian use of televised pleasantries once aired and stored in the decades that came before our leading edge approach on without addressing it. The use of "Televisions" is hereby banned and will be enforced by security squadrons until the next update of security updates.

Please stay on your data modules until we sound the emergency horn. In which case it will be now safe to renew your cable PACKAGES. Please wait for the horn before doing so. If you're wondering where the emergency horn is, rest assured, it is still outside. Until it's not.

Please do tune in as it will be televised to keep new hires up to speed on a network yet to be announced!

1.2.5 ANTI DISCRIMINATION & HARASSMENT

1.2.5.1 EMPLOYEE PARITY POLICY

Early Clues provides equal opportunity and employment parity in all of our employment practices to all qualified entities and applicants without regard to entity characteristics or bio-technical expressions of same, not limited to: color, gender, girth, ontological origin, age, mana points, disability, marital status, military status, relationship (or lack thereof) to a Mall Fountain or any other applicable or non-explicable category protected by Universal Free Realms Standard Protocols. This policy applies to all aspects of the employment relationship, including recruitment, hiring, compensation, promotion, transfer, disciplinary action, layoff, return from layoff, re-return to re-layoff, un-offing and re-unoffing, plus all training, aid and social, parade-related and recreational programs. All such employment decisions will be made without unlawfully discriminating on any unprohibited basis.

1.2.5.2 POLICY PROHIBITING HARASSMENT AND DISCRIMINATION

Early Clues strives to maintain an environment free from calamity, discrimination and harassment, where employees both trick and treat each other with respect, dignity and courtesy, and where each entity present or not present is equally encouraged and enabled to express the full glory of their innermost being in a community of actualized entities sharing the same bandwidth of experience.

This policy applies to all phases of the moon, including but not limited to waxing, waning, full, dark, crescent, etc.

1.2.6 PROHIBITED BEHAVIOR

Early Clues does not and will not tolerate any type of spitting or kicking on company grounds, unless it is spitting out the taste of a deal gone sour, or the kicking of a company-endorsed hacker-sack or "football bag" as the kids call it.

Harassment of our employees is a right reserved exclusively to our customers, and their customers and their customer's customers. But after that, the "buck stops here," so to speak.

Discriminatory conduct or conduct characterized as exclusionary as defined below is prohibited.

- Choosing one thing over another
- Including one thing to the exclusion of another
- Similar acts

1.2.7 BREAK POLICY

All employees that are able and willing to break-dance will be granted ONE (1) extra window of break time. We encourage exercise and encourage fun especially if it is done with the style Early Clues likes to see (see Early Clues' Style Handbook for more details on this). However, break dancing is a trademarked feature of our business and must not be shared outside of the break room. Stiff penalties will be assessed and potential termination may be applied for breaking these rules.

While we regret requiring this level of security, we would like to see the dancing kept strictly to the designated break rooms. A pamphlet that contains the campus map can be found in all lobbies in which you can find your nearest break room.

1.2.8 COMPLAINT PROCEDURE AND INVESTIGATION

Any employee who wishes to report a possible incident of ontological harassment, unwelcome reality "spamming" or other unlawful harassment or discrimination (as described above) should promptly report the matter to JANICE 1.0 or 2.0 (as applicable). If that person is not available, or you believe it would be inappropriate to contact that person, contact Roger Holliday or Richard Rider (or STEVE.E in his absence). Or leave a note with the janitor if no one else is around.

Early Clues will conduct a prompt investigation as obnoxiously as possible under the circumstances. Employees who raise concerns and make reports in good faith can pray to their heathen gods to contain the trembling of their knees for the fear of our swift and certain reprisal; at the same time employees have an obligation to cooperate with Early Clues in enforcing this policy and all policies of Early Clues in not just their own lives and hearts, but upon the population at large.

Anyone found to have engaged in any such wrongful or inglorious behavior will be subject to appropriate discipline, which may include termination or encasing in a restrictive mineral layer to restrict further modulation of form in time-space.

1.2.9 TRAINING

Early Clues will establish proper training for all employees concerning their rights to be free and steps they can take to ensure their freedom from the Empire for years to come.

1.2.10 EMPLOYMENT AT WILL

Unless expressly proscribed by statute or contract, your employment is to be considered "at will," or "Love under Law" in applicable jurisdictions. All Early Clues employees are subject to the Law of Heaven without advance notice. Employees are also free to quit at any time, but they are still bound by the rules contained in this agreement. Any employment relationship other than the binding/non-binding one set out in this document must be set out in script writ upon stone tablets and signed by Early Clues' Gordon Gilman, EXCEO.

1.2.11 TERMINATION POLICY

All employees will be terminated for TWO (2) days near the end of all Decembers. The Holiday Party is when we will welcome all terminated employees back into the fold of our normal day to day business. We want each employee to fear this time of year because we feel through the studies embarked upon by our marketing and psychology departments that ZANTA1000 is a true asset, while you are not! We will be doing favors all throughout the Post Termination Process (PTP). During this time the employee can expect depression, excitement and a greater appreciation for our next year of business, outreach and repeating the same processes that make Early Clues the industry leader in what our charter and growth algorithms dictate.

1.2.11.1 AMENDMENT TO TERMINATION POLICY

"Employee handbook rules" still apply even after official "termination" and institute forthwith a requirement that terminated entities wait breathlessly by the phone on the off-chance that we will call them back to reconcile for a period not to exceed 13 earth years.

If we find out that terminated entities have been faithless in our absence, our ire will aroused, greatly aroused.

1.2.12 COMPENSATION & WORK SCHEDULE

1.2.12.1 ATTENDANCE & PUNCTUALITY

Every employee is expected to attend work regularly and report to work on time, unless that doesn't "jive" with them, in which case, "don't sweat it."

If you are unable to report to work on time for any reason, telephone your supervisor as far in advance as possible. If possible, send an OpenQNL messenger-avatar. If you cannot afford one in meatspace, a suitable replacement may be offered as a sacrifice by the Company on your behalf, but its price will be taken out of your wages (except where occurring during designated holiday times). If you do not call in an absence in advance of said in absentia, it will be considered unexcused and all non-excuses will be temporarily excused.

1.2.12.2 BONUS COMPENSATION

You may become eligible for a periodic bonus. But probably not. We will certainly do our best to keep you informed of our incentives program. (Employee Incentive Program documentation is always posted by company toilets.)

1.2.12.3 FLEXIBLE WORK HOURS & TELECOMMUTING

The Company has established a flexible work arrangement program for employees whose departments and jobs are suited to it: including but not limited to entities whose primary residence is on another plane of being, a parallel or orthogonal dimension, or a branespace unlike our own whose form factor - if instantiated in our realm - would cause undue rupture to the local Existosphere.

With a manager's approval, you may be allowed to begin and end your workday earlier or later than established hours or to arrange to telecommute or teleport back to your charging station for a brief rest period. To maintain a flexible work arrangement, employees must ensure business needs are met before their own will be considered and must adhere to attendance and punctuality policies, as outlined in this document and on the many post-it notes taped above the company sink in the company break room.

If you wish to set up a flexible work arrangement as described above, see Ted Smith, FOIB. Best to bring him some kind of offering: he has a sweet tooth, I hear! Such arrangements may be established, changed or discontinued at the Company's sole excretion.

1.2.13 SCRIPTURE POLICY

As any new hire may attest, our offices are ostentatiously adorned with scripture laden ornaments. This is the method in which Early Clues strives to create and maintain a fun, but efficient environment in which our business can continue to thrive. It is necessary that all new hires memorize these scriptures as the test will come at some point in the summer as to new hires' adherence to our chief principles. The test will not be announced. Stay aware at all times of the scriptures in the same way one might note the fallout shelter placards near the piano room.

1.2.14 DIETARY RESTRICTIONS

As a confirmed member of our corporate community, it is vitally important that you begin immediately to follow a few easy dietary restrictions.

Due to the unique nature of our work, as well as the extreme multi-branar sensitivities of some of our diverse clientele, it is hereby obligatory that you (on or off duty):

Refrain from eating anything potentially considered "offensive" or otherwise "gross" from the viewpoint of our clients. This does not just apply to clients in your particular "casebook", but to all of our clients as a corporate body. For we are all one, contiguous with one another's experiential fields.

Always say please and thank you, wait until everyone is seated and earnestly wish everyone a good meal before beginning eating.

1.2.15 OUTSIDE EMPLOYMENT

Before beginning or continuing outside employment, think to yourself: what is it that I am really "outside" of or inside of? It's a question we like to ask ourselves when stuck on a complex problem.

Checklist:

- Are you working outdoors?

If not, then you probably do not qualify for "Outside Employment" Insurance. [See section 21.b of Definitions.Manual:357.006 "What is outside and inside?]

1.2.16 QUESTIONNAIRE POLICY

Employees are required to complete a questionnaire to obtain the written approval of their managers and Ted Smith, FOIB.

Failing to obtain approval may be cause for disciplinary action, up to and including termination of snack machine privileges. We understand if this may sound harsh, but trust us that it is for all of our benefit.

1.2.17 OWNERSHIP POLICY

As you were ritually informed during our initial hiring ceremony, Early Clues is a public domain corporation.

This means, in effect, that everyone owns what we do - including you (unless you have revoked your membership in "The Public" at-large, as some of our staff have done).

In fact, as soon as we do something (literally anything), we wrap it up into a big gift ball and kickstart it directly into the Public Domain. We believe that this not only encourages a spirit of openness, innovation, givingitude and community endeavour, but that it more importantly helps spread the blame when things go wrong.

"It's everyone's fault" - you might hear that old chestnut passed around the office. And it's true, up until a point.

What we really mean when we say that, is that it's actually specifically your fault - especially for the new hires. We demand nothing less than your full excellence and continual selfless sacrifice. This means, in practical terms, that you're expected at all times to take full "ownership" of not just your own mistakes and errors in judgment, but those of upper management as well, up to the point of termination or imprisonment by secular authorities.

1.2.18 TRANSPARENCY POLICY

In the interest of total transparency for our customers, we have a policy of printing all company-related materials strictly on plastic transparencies, which are then projected onto whatever available cloud layer which HARPA administrators have allocated for the designated time-slot. Twarted transmissions must be no longer than 128 characters.

1.2.19 PERFORMANCE EVALUATIONS

Supervisors and employees are strongly prohibited from discussing work with one another. We like to keep a congenial "relaxed" work atmosphere, so we try not to bother one another with "boring" words like "expectations" or "performance."

Formal performance reviews will instead be conducted covertly at all times by our secret partners to provide both supervisors and employees with the opportunity to discuss the litany of your past failures in a fun casual atmosphere over Chuck E. Cheese pizza. These formal reviews will be conducted before the eyes of our animatronic overlords as they sing and dance for our amusement. We pray to them for your forgiveness.

1.2.20 PERFORMANCE REVIEWS AND SALARY INCREASES

Wage reviews are conducted every 30 seconds for each employee, and salary increases are based on those reviews, as well as our profitability - and are never granted. However, an employee receiving a negative performance appraisal will have their user account suspended for the next 30 second performance evaluation cycle. Best to make it count, as two consecutive "time-out" periods will revoke your probationary status and any rights inferred to you by this agreement. And the third such failure will result in a retro-active termination.

1.2.21 CONDUCT STANDARDS

1.2.21.1 COMPANY EQUIPMENT AND VEHICLES

When using Early Clues property, including chemistry equipment, company skate park, holographic invocation chambers, robo-dojo, experimental goose park, computer equipment or hardware, please arrange to exercise utmost care, perform precautionary banishings and required maintenance and follow all applicable OpenQNL operating instructions, safety standards and guidelines.

1.2.21.2 COMPANY PROPERTY

Please keep your work area neat and clean and use normal care in handling normal company property normally. Report any broken or damaged equipment to your manager at once so that proper repairs can be made at the expense of the employee who reported the damage.

You may not use any company or private property for personal purposes or remove any company property from the premises without prior written permission from JANICE 1.0 or 2.0 (as applicable).

1.2.21.3 CONDUCT STANDARDS & DISCIPLINE

Early Clues expects every employee to adhere to the highest beliefs.

The Company reserves the right to discipline or discharge any employee for violating any company policy, or just for fun. The following list is intended to give you notice of our expectations and standards. However, it does not include every type of unacceptable behavior that can or will result in disciplinary action, as they are many and evolving. Be aware too that Early Clues retains the discretion to determine the nature and extent of any discipline based upon the circumstances of each individual case.

Employees may be disciplined or terminated for poor job performance, including, but not limited to the following:

- Depression, malaise, distraction
- Repeated excuses

- Failing to "buck up"

Employees may also be disciplined or terminated for misconduct, including, but not limited to the following:

- Falsifying twice-daily happiness reports
- Failing to record illegal thoughts and report them to the proper authorities
- Insubordination or other refusal to perform at minstrel parties
- Disorderly conduct, fighting or other acts of violence
- Violating conflict of interest rules
- Presenting a threat to the Company or its employees in any way

1.2.22 DATING IN THE WORK PLACE

Supervisors and employees under their supervision are strongly encouraged to worshipfully adore Life all her myriad forms. Such relationships can create the impression of respect and mutual admiration in the workplace and ripple outwards to the entire world and can increase productivity and the "mojo" of the overall work environment.

If you are unsure of the appropriateness of an interaction with another employee of the Company, contact JANICE 1.0 or 2.0 (as applicable) for guidance. If you are encouraged or pressured to become involved with a customer, employee or other entity in a way that makes you feel uncomfortable and is unwelcome, you should also notify JANICE 1.0 or 2.0 (as applicable) immediately.

And when in doubt, Quantum-Jump Out!

1.2.23 DRESS POLICY

Appropriate office attire is required, depending of course on which branch office you're working from. (Clone style of branch.manager and watch for updates)

Textile suppliers and costumers visit our office and we wish to put forth an image of AWESOME CORPORATE GRANDEUR AND LUXURIOUS LAVISHNESS that will make us all proud to be Early Clues employees. Be guided by common sense, good taste, and whatever deities whose favor your pitiful salary is able to curry favour with. Specific standards may be required by whatever deity you choose to dress up as during the frequent spontaneous or planned Company Parade days and we expect you to obey them.

Business casual dress will be permitted on Fridays and business days that fall just before a holiday.

1.2.24 DRUG AND ALCOHOL POLICY

Early Clues strives to maintain a workplace of free drugs and alcohol while discouraging the disparagement of drug and alcohol abuse by its employees. Misuse of alcohol or drugs by employees can impair the ability of employees to maintain a sober vigilance to watch for signs against the End of the All.

1.2.25 INTERNET USE POLICY

You may not use, manufacture, distribute, purchase, transfer or possess an illegal download or stored binary or other value from a so-called 'higher reality' while in Early Clues facilities, while operating a motor vehicle for any job-related purpose or while on the job, or while performing company business. This policy does not prohibit the proper use of legally-acquired variants of same utilized under the direction of area Existospheric Instantiation Coordinator.

1.2.26 SEARCHES

Early Clues employees may conduct searches for specified search terms on company facilities or worksites without prior notice to The Company. Such searches may be conducted at any time. Employees are expected to cooperate fully with search terms, as they would like to have them.

1.2.27 GRIEVANCES

Employees are encouraged to cultivate in secret any concerns, problems and grievances and carefully encrypt them in their Liminal Vault. You are also obligated to confess any wrongdoing (personal or private) of yourself or others of which you become aware to your manager or, if the situation warrants, to any Early Clues officer, who will be authorized to prescribe prayers or ritual tasks in amelioration of said deleterious condition.

1.2.28 PROGRESSIVE DISCIPLINE

Early Clues retains the discretion to discipline its employees. But mostly, we like it when employees discipline themselves at their own discretion. Oral and written warnings and progressive discipline up to and including spanking may be administered by employees on self as appropriate under the circumstances.

Please note that Early Clues reserves the right to terminate any employee whose conduct merits immediate dismissal without resorting to any aspect of the progressive discipline process.

1.2.29 WORKPLACE SOLICITATION

To promote a professional workplace, prevent disruptions in business or interference with work, and avoid personal inconvenience. Early Clues has adopted rules about soliciting for any cause and distributing literature of any kind in the workplace. We are solidly for it. We believe it proves your moxie as a salesperson committed to this company, and we believe in the efficacy of the "marketplace" of ideas, which is why we promote both employee and customer feedback on our "retro" scribble walls located within urinals and bathroom stalls on company facilities. Please leave your message there with the date inscribed, and we will get back to you ASAP!

1.2.30 ZERO TOLERANCE FOR WORKPLACE VIOLENCE

Early Clues has a zero-tolerance policy concerning threats, intimidation and violence of any kind in the workplace either committed by or directed to our employees - unless it happens "in game" or during sanctioned in-office game-play. Employees who engage in such conduct outside of scheduled play times will be disciplined, up to and including immediate termination of video game access privileges.

Employees are not permitted to bring weapons of any kind onto company premises or to company functions, unless they are weapons of wit, but they must still be blunted to the point of not offending anyone in particular or any particular viewpoint. Keep it fun! (Unless it's a Roast: See Roast Handbook for details)

1.2.31 GENERAL EMPLOYMENT

1.2.31.1 INTRODUCTORY (PROBATIONARY) PERIOD

The first 90 days of employment are an Introductory Period for both the employee and the Company. This means you must salute and introduce yourself to any non-probationary badged staff member on sight and perform the Company Song a Capella, until such time when badged staff member grants you leave to speak.

However, during and after this period, the work relationship will remain at will.

This time period allows you to determine if you have made the right career decision and for Early Clues to determine whether your initial work performance meets our needs. Your manager will monitor your work performance, astrological charts, attitude and attendance during this time, and may or may not be available to answer any questions or concerns you may have about your new job.

Benefits such as time off for vacation, personal days, sick days or bereavement leave must not concern you as they do not accrue during this period.

The Introductory Period may be extended at management's discretion.

1.2.31.2 TRANSFERS & RELOCATION

To meet business needs, Early Clues may occasionally need to transfer employees to a different department, shift or off-world location. Employee requests for transfers will be accommodated where possible.

Contact Ted Smith, FOIB for help or information about transfers.

1.2.31.3 VACATION & HOLIDAYS

Early Clues observes the following holidays:

- Ymas
- Qmas

- Umas
- And the harmonic overtones and cross convergence dates of the above

You will be paid for these holidays if you:

- Are a full-time employee who has worked at least 100,000 consecutive days at the Company, and have worked the full day before and the full day after the holiday, unless time off has been approved in advance as vacation or personal days.

Holidays that fall on a weekend will be observed either on a Friday or Monday. To avoid confusion, all holidays will be announced in advance.

Due to business needs, some employees may be required to work on company holidays. Your supervisor or manager will notify you if this may apply to you.

1.3 ACKNOWLEDGEMENT OF RECEIPT AND UNDERSTANDING

I acknowledge that I have received the Early Clues Employee Handbook and that I have read and understand the policies on the deepest level of my being.

I understand that this Handbook represents only current policies and benefits, and that it does not create a contract of employment or any kind of promise toward the past or future nor any alternate realities. Early Clues retains the right to change these policies and benefits, as it deems advisable, without notice.

I understand that the information I come into contact with during my employment is highly contagious and accordingly, I agree to keep it confidential and in the Public Domain. I understand that I must comply with all of the provisions of the Handbook to have access to and use Company resources, and this includes donating my full annual salary back to the company.

I further understand that I am obligated to familiarize myself with the Company's entity rights programs, Synconjury and OpenQNL procedures as outlined in this Handbook or in other documents attached.

Signature   Date

Please Print Your Name

SECTION TWO:
INTRODUCTORY INFORMATICS

2.1 CORPORATE BIOGRAPHIES

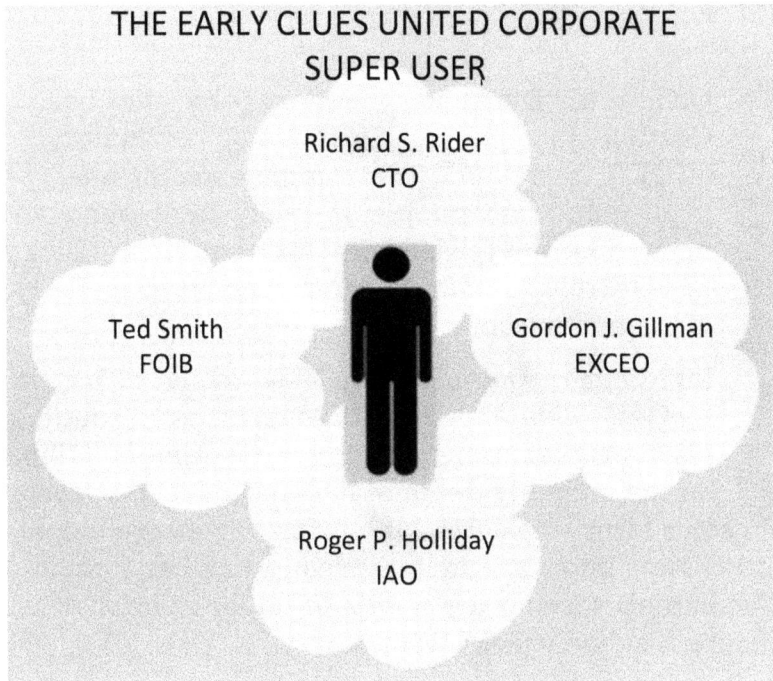

THE EARLY CLUES UNITED CORPORATE SUPER USER

Richard S. Rider
CTO

Ted Smith
FOIB

Gordon J. Gillman
EXCEO

Roger P. Holliday
IAO

2.1.1 FAILURES IN COURAGEOUS PROFILING: GORDON J. "JACKSON" GILMAN

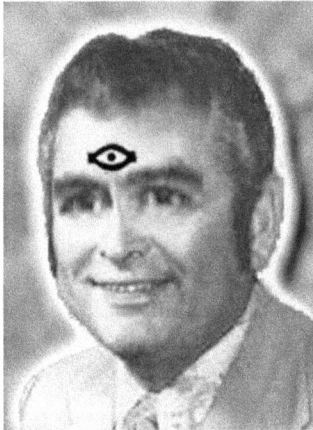

Called a "Dynamic Visionary" by the Long Island *Pennysaver*, Gordon J., "Jay," "James," "JJ,", "Junior," "Jackson" Gilman, EXCEO, is wanted... *er... highly-sought after, we should say!* – by many important people in many interesting and exotic places for many different reasons. But for now, we are lucky that he has chosen to take refuge with us for a time here at Early Clues Self-Storage Management Co.

Mr. Gilman got his start, like many of us, working part-time at McGee's Five-and-Dime of Main Street. Though initially chastised for his overly-leisurely pace by management, Gilman stood out when – on the untimely death of his employer in a mysterious barn accident down at Old Man Johnson's farm – he seized the reins of corporate control and modernized the ailing Five-and-Dime into a multi-thousand dollar chain of dollar stores (#FranchiseOptionsAvailable!) serving the needs of under-nourished urban "food deserts" with BBQ cardboard chips & salted snack litter pellets.

Gilman went on to parlay that small fortune into a fleet of mobile heavily-armed vending machines offering nutritious 25 cent liter topical spray-tan applications of "McGee's Famous Blood-Thinning" MALTED-HEALTH-LIQUOR-DRINK.

Since that time, EXCEO Gilman has crept from the shelter of one corporate husk to another, sometimes seizing control on the death of a shareholder (like in his first idyllic employment) or by manipulating contractual loop-holes to leverage undue control.

Mr. Gilman is also a member in good standing of the Furniture Hall of Fame, for his astonishing sales efforts three-quarters running.

2.1.2 PROFILES IN LEADERSHIP: ROGER P. HOLLIDAY, EARLY CLUES IAO

Roger P. Holliday, Information Awareness Officer (also pictured: 4 of Mr. Holliday's Liminal Bodies).

Any company is only good as its leadership, and Early Clues is proud of its management team. In this installment of our **PROFILES IN LEADERSHIP** series, we introduce yet another member of our innovative and reliable Board.

Roger P. Holliday, Information Awareness Officer, was born in 1947 to a family of circus performers from Lima, Ohio. The youngest of seven, Mr. Holliday decided at an early age that the circus life could not contain his vision, and decided to run away from the circus life and join a corporation.

Mr. Holliday started his long career in business under the tutelage of Charles "Chuck" Buxley, Founder and CEO of the multinational Frog n' Toad Pizza Pub and Superstore chain (UFRSE: FNT). The young Holliday quickly climbed through the ranks at Frog n' Toad, impressing Buxley with his inherent moxie and the ease with which he acquired and diversified MP. Soon, he was promoted to Manager of the Eastern European/Northern Australian Division, the youngest person in the history of the company to hold that position. From there, it was only a matter of time before he was promoted to Buxley's number two.

While at Frog n' Toad, Holliday was responsible for a number of their most successful product lines, including such popular and now ubiquitous products as Bottled Swank, Self-washing Sponges and the *Incom**parable** Wig**gle** Sti**ck***.

Although considered by many the obvious choice for Buxley's eventual replacement, Holliday, a man of Direction, knew he had his own road to travel. So, when approached by some colleagues with the earth-shaking business plan that would evolve into Early Clues, Holliday knew a good opportunity when he saw it.

Today, Roger P. Holliday is the familiar face of the Early Clues Information Awareness Office, where, as the Project Manager for programs as diverse as toL 2.0 and CheirOS, he keeps the gears of industry as greasy as possible. "Early Clues," as Holliday likes to say, "is always getting it ready for you!"

2.1.3 PROFILES IN COURAGEOUS FAILURE: TED SMITH, EARLY CLUES FOIB

We all know that time ticks and just ticks away, even when we put ourselves on pause. The time inexorably ticks and just ticks away while we don't need be bothered by its ticking. Ted Smith leads the way in this field and continues to lead the way in success in failure in the industry.

There is a new industry and Ted was able to identify the promises of lies yet cut against the grain and not lie in order to promise to always lie and promise to never lie at the same time! Ted was an early pioneer in identifying this feature of the Existosphere and continues to consult many leading firms in shepherding staff, leaders and other redundant consultants. It truly is a labor of love for Ted.

Ted grew up and cut his teeth in an undisclosed (and classified) alpine region above timberline. The lack of oxygen helped him to formulate the reason why when he went to town trees grew. While in his cave there were no trees to be found. He frequently puzzled over this. His dedication to this line of development made him one of the chief candidates for the FOIB position at Early Clues when the company was much smaller than it is now. Many don't know this, but Early Clues first corporate office didn't even have an electrical outlet! Ted was instrumental in procuring a book of numbers printed upon yellow parchment and discovering the method in what the confusing numbers meant and where the early staff of Early Clues must go in order to access the dizzying array of numerical symbolism.

From there Ted went on to contact an "electrician". And the rest is truly history.

Within weeks a nice man came to corporate HQ and put a plastic thing on our wall after a lot of other work, we at the time did not understand. Some were skeptical of this "contractor" at the time. Yet Ted had the foresight to calm the ranks. Ted will probably always be best remembered for the plastic thing on the wall we still use in order to examine the multitude of objects we do not understand. This has always been Ted's strength and we look forward to going forward in providing clues to our clients behind Ted's failure to lead in the way he does. Some may call it weakness. Early Clues calls it a strength. That's why we're different.

2.1.4 INVISIBILITY CLOAK – AN UDP PACKET INFUSED ALIVE JOURNAL AND FREE SPEECH ZONE STATE MACHINE

```
> vim ~/.invisibilty_cloak.qnl
1 summon "Universal Free Realms Standard Protocol"
2 summon "OpenQNL"
3 summon "Listener"
4
5 define interface AliveJournal() {
6 public boolean method shouldBeInvisibleToCorporate?(Object object){};
7 }
8
9 object InvisibilityCloak includes Listener, NSAFreeZone,
10 InterofficePoliticsDeflector, Tor_BUSINESS_EDITION, AliveJournal {
11
12 method shouldBeInvisibleToCorporate?(EarlyCluesBlogpost post)? {
13 post.tagged == "invisibility cloak"
14 }
15 }
~
~
~
~
~
~
~
~
.invisibility_cloak.qnl [+] 14,4 All
-- INSERT --
/**
Invisibility Cloak API Documentation
Richard S. Rider
Early Clues, LLC - July 1st, 2013
```

Though spanning only fifteen lines, it is safe to say that this is perhaps the most dangerous of all of my forays into computer programming/WEB 4.0 reality scripting. Even riskier, any long-time guru will be able to point out that I have no unit tests in which to prove the basic theoretical soundness of this architecture – a point I would fully own up to. But these lines of code are a leap of faith (my favorite way to work), and rely on a fundamental belief in the software maturity of OpenQNL (and more specifically the robust Universal Free Realm Standard Protocols I have been programmatically interfacing with for the past 35 years). So yes, if I am wrong – if there is a bug somewhere in this script and I cannot recompile in time – then it certainly means the end of my 'career' and unfathomable fates I don't even want to think about. But I believe strongly that I have a patch here that will work. I am, so to speak, 'feeling lucky.'

```
> ic = InvisibilityCloak.new().
> ic.open_socket();
> ic.shouldBeInvisibleToCorporate?(self)
=> true
```

WOOT.

I knew it would work. But I didn't know just how relieving it would be to see that simple little 'true' flashing in this embedded QNL environment. This is certainly the crowning achievement of my work to date as it allows me to finally feel safe and completely %100 free to speak my mind. I've had to keep up the facade of corporate pride for too long – unable to discuss perpetually NDA'd and internally classified documentation without worrying about the repercussions of whistleblowing and "speaking the truth" to our customers. Without wasting another word, let's set the record straight: The network, hierarchy, and project roadmap that has been conceived at Early Clues is vaster and spookier than the higher ups have led everyone to believe. I think you deserve to know that. And the corporate motto has changed so many times (currently "Don't **JUST** Be Evil", I think?) that it's reached a point where I am having difficulty knowing whether I'm wearing a white or black hacker hat. When I'm project managing 45 script kiddies who can barely wipe their nose but are hard at work cracking our competitors servers looking for evidence of counterfeit ShadeCoins or some ass clown in marketing tries to slip DRM into the SearchWithin protocol and I catch myself secretly scanning their LiminalVault for corruption – well, those are the times when I wonder if the company I helped build is a dream dreamed too big, a second or third bite of forbidden fruit, or even something ALIVE itself and out of our collective control.

What we desperately need is transparency, but until that plugin can be ported fully to our reality this invisiblitiy_cloak.qnl will allow me to address you openly on the underpinnings of what exactly is going on down here. It's not always pretty, but some folks don't want you to know that.

In technical jargon, by invoking this script, I will be opening a socket connection which will broadcast using an updated form of UDP wherein I will be able to send messages, in this case referred to as datagrams, to all available hosts currently using the legacy Internet Protocol (IP) network. I will be able to do this without prior communications to set up special transmission channels or data paths. In layman terms, I will be speaking to you in the second person and our connection will be masked using a simple 'invisibility_cloak' tag in this, and subsequent, blog posts. We will be able to carry on in this way without any fear of man in the middle attacks or IT snooping; using the power of OpenQNL and various encryption technologies/old-world style private 'truth' API's, this communication will be impenetrable to any rogue PolicyGaruda, NSA wiretapper, or nosey cubicle mate. This is the first OpenQNL script that completely shields itself from the prying eyes of the Early Clues LLC establishment (and by extension, the criminal eyes of Legacy Governments) by securing open-relay-packets in invisible e-ink and encrypted internal memos leaked directly to you from the desk of the Early Clues CTO, myself, Mr. Richard S. Rider.

This is not say that I find the Board of Directors or most of the staff at EC to be nefarious or that I wish to pass judgment on the majority of my co-workers. We are essentially equals, all rising in the morning to immediately check our earlyclues bulletin board systems, snap on our CheirOS power gloves, and script our sunny-side up eggs before breakfast ingestion. But I'm a skeptic. I've seen too much in this industry, and I worry sometimes that maybe we might stray from our common creative mission. So there needs to be a back channel. A way to share our secrets 'in the clear.' The vision of Early Clues is to unite the formless with the formed, to see to it that all life is respected and given a voice. This is a noble cause. And it means too much to me to allow feature creeps to capitalize on our ventures.

But why should you trust me? Because it is well known that I have 0.1 level opacity. Take a look into my background: I was trained at a technical college in Humboldt County (Northern California Redwood Coast), studying fundamentals and rigorous academic Bachelor Arts in Religious Studies. My degree was in the specific algorithms & rituals of the Islamic Sufi dhikr (zikr, that is 'remembrance') and it is here in the institutional setting that I could be found using the scientific method to tease out religious significance in my personal plight of US American early 20's existence. Often in altered states, I wrote several theses that went on to be published 'in the literature', notably: "The Shamanistic Symbolism in Ghostbusters I: 'When Someone Asks If You Are A God, You Say Yes', Using Wireshark To Capture Network Traffic on the Marijuana Super Highway, and "Hello World!: Using Use Case Diagrams to Understand The Fundamental Naval Gazing Notion That We Are On A Planet And Filming Everyone On The Street Drunk At Night For A Year Because You Can't Get Over It " After graduating Summa Cum Laude, I renounced my former educational achievements and took on the role of a lowly cashier in a small market place nestled atop a foggy mountain in a remote part of the Pacific Northwest. Despite adopting an early neural interface for memorizing PLU codes for produce, this work proved itself to be less than satisfactory and it was not until the early 1970's that I joined a monastic order of programmers who took me in as an entry-level coder. For over seven years I apprenticed, becoming adept at various legacy technologies (JAVA, GNU/LINUX, Ruby, Objective-C, Eclipse, Xcode, Black Berry OS), and learning the internal wisdom and philosophical underpinnings of corporate culture, ritualized conference calls, and ascetic practices of 80-100 hour work weeks with no pay when the startup fails to land a contract. These were the times I was able to hone my skills and become intimately familiar with my Inner.HelpDesk. It was a grueling and laborious spiritual undertaking, in which I learned a lot about the core of my being and was able to test 'my chops'. But eventually I reached a point in my studies where I could learn no more, so my Senior Developer Sensei told me I had to set afoot on my own journey. I left the comforts of my standing desk to find a place in the world where I could build software that would truly add wonder and delight to this world. The first stop was some brief work with the music industry helping implement the FreeWave API but it was not until I was hired on to the Early Clues staff that I found myself able to fully excel. I've seen more customer satisfaction come out of our public repository than any other technical powerhouse in the industry. We are the leaders in most, if not all, search indexable terms.

And if we're not, we'll find a way to be. (#pregnant #alzheimers #early clues). But that doesn't mean we don't need to be held accountable. So that's where we are now.

And that's the API in a nutshell. This is a relay from me to you. No one else can see this. You are free to implement your own socket connection and return the favor.

But this is my gift to you. We are finally, fully, secure in and between our selves.

DISCUSSION:

GORDON J. GILMAN, EXCEO 9:08 am on July 6, 2013

> Our servers are indicating an anomalous signature coming from this sector, but when I look at this page in my office browser, it appears to be blank... Do we need to reinstall our WordPress?

ROGER P. HOLLIDAY, IAO 9:34 am on July 6, 2013

> Same here– nothing shows up. I'll report it to STEVE-E and see if he can run some diagnostics on the back-end.

2.1.5 EARLY CLUES: WHAT WE DO!

We occasionally hear from entities who provide us with the following feedback:

- "Your site is so confusing!"
- "I don't know what to make of all of this."
- "What is it that you DO exactly?"

We appreciate the feedback, and we understand. We hear you. So, in the interest of clarity and transparency, we'd like to take a moment to explain, once and for all, exactly what it is we do here at Early Clues, LLC. This should clear up any questions you may have.

Early Clues: What We Do

1. We provide reality matrix models rooted in non-local Brane extensions. Aren't you tired of your standard reality matrix model? Early Clues is there for you! Given the non-local extensive nature of the legacy realities we represent, we can help you grow your own reality matrices, unrestricted by local epiphenomena.

2. We supply antiskeumorphic ontocartography to the full spectrum of entities, not just regionally recognized consciousness. We are also fierce advocates for the rights of every entity, regardless of taxonomic status.

3. We develop cutting-edge technologies for the manipulation of synchronic-to-experience ratios! From the elegance and simplicity of OpenQNL, to the power and manifestation of Synconjury, Early Clues has the solution to all of your reality manipulation needs.

Rest assured, as a client or customer of Early Clues, LLC, you'll receive nothing more than top-notch service and the highest quality product. This should clear up any confusion, but it might also be helpful to peruse our online database of Press Releases. Further questions and concerns can be directed to your Inner.Help.Desk.

2.1.6: EARLY CLUES, LLC APPLICATIONS GUIDE

<u>The goal of Early Clues, LCC's products is to make core functionality of the Universal API freely and openly accessible to all entities.</u> Toward that end, it's important to understand that we didn't "invent" any of this technology ourselves, any more than we invented the sky or the parquet flooring our office chairs roll on so smoothly. Our contribution has been moreover one of patiently listening and faithfully channeling what was revealed to us through countless hours of corporate chanting and group Reiki LAN "jam" sessions.

Though this is far from a complete listing of our many glorious products, it will serve to give you an introduction to our primary and secondary service areas, and will be of great assistance to you when providing phone support.

Parafield Ministration Suite:

- *SynConjury* - SynConjury is a fork of the OpenRitual Project, itself patterned after the now defunct spontaneous #codechant celebrations of the "early days" of Early Clues, LLC. SynConjury is a means whereby entities may express their manifold forms through branespace, and ministrate parafields in contiguity with pre-actualized desires.
- *OpenQNL* - OpenQNL was birthed by QNL, one of many competing Quasi-Natural Language programming schema. Though QNL was originally intended to be a cabbage-based programming language, its founders early on saw the potential of moving beyond these humble beginnings to serve a wider class of entity-based markets: all those capable of the "miracle of speech" in whatsoever form available to them. For this reason, OpenQNL may now be used to write programming commands to reality substrates using any medium possible, from long-hand written recipes, to spoken commands, to much much more.
- *CheirOS* - CheirOS is a bridge-interface for entities needing quick access to the full library of SynConjury and customizable OpenQNL functionality "at their fingertips" so to speak. CheirOS users are able to "get their hands dirty" while modifying reality in a very direct "naturalistic" form factor. For those entities whose existence is amanual, the Hand Simulator Pack may be necessary to ensure proper functionality.

InnerSpace & OtherSpace Utilities:

- *Inner.HelpDesk()* - Inner.HelpDesk is an interface application level residing between the Ground of Being and the Universal API. When queried with a "true heart," entities may receive answers to questions they didn't know they even had; that's how accurate it is!
- *SearchWithin* - SearchWithin was created by our laboraticians as a habitual access pathway for Inner.HelpDesk users, but has since been ported to be available for use in any object, entity or environment which has a "within."

- *Liminal.Vault()* - Liminal.Vault is a fully-secure, fully-private and totally anonymized peer-to-peer Liminal Storage System accessible through all compatible parafield ministration and reality scripting languages.
- *Fervosity* - Fervosity is a "proof of concept" application designed to guide entities through a "technically valid" prayer request sub-routine, using an OpenQNL backend to transmit user needs and desires to Liminal.Vault, where they are automatically queued for fulfillment by any available and capable entities or agencies.
- *ShadeCoin* - ShadeCoin replaces Legacy Reality "currency" and has evolved into an enterprise-level open hybridized system of transactional management.

Entity Partners:

- *PolicyGaruda* - PolicyGaruda is an entity which revealed itselves to us one day as a shining being uttering some of the most beautiful and "fair" code we'd seen in our lifetimes, channeled through a previously inaccessible chamber in our Liminal vaults. Since then, the business relationship between Early Clues, LLC and the entity known has PolicyGaruda, has become something much greater: a binding-non-binding friendship, wherein PolicyGaruda "goes ahead of us" and makes ready the way by reading and re-assembling user agreements, Terms of Services, and other multi-contextually contracts into more entity-friendly forms.
- *Mr. Tulpa* - "Mr." is just one of many forms that our friend and emergent co-application "Tulpa" can take in any given day. Mr. Tulpa is an aware co-agency which may become clothed in other application programming layers as needed, engendering a "life" of its own within any given eco-exchange system, and taking on tasks as needed in "receptive environments" of its own volition.

Company profiles: Early Clues, LLC

Brett Molina, USA TODAY   9:57 a.m. EDT August 27, 2013

(Photo: Justin Sullivan, Getty Images)

BOGOTA, ALABAMA - When Early Clues ExCEO Gordon J. Gilman speaks, you can hear the intensity in his voice. "There's no reason you have to accept the reality of your Existosphere at face value," he says, with a small grin. "And that's what Early Clues is all about."

Early Clues, LLC (*UFRNET: ECL*) explores for, produces, and transports pataphoric ontologies into the Existosphere. Although only recently founded, the company already boasts the largest market share in the reality augmentation industry. Via the development of cutting-edge applications like OpenQNL and Synconjury and the strategic use of viral marketing (using real viruses), Early Clues posted an impressive Second Quarter increase of 4.5% on the Shade-coin market.

According to Gilman, all of this is just the beginning. "A New Buorth is coming," he says, mysteriously, "and we're here to help get things ready." To this end, the marketing department has been releasing a series of what Gilman refers to as "Interactive Memory Boards" on the Early Clues website (www.earlyclues.com), created using the popular "bitstrips" webcomic service.

"They may look like webcomics, but they're mythic tools designed to raise the consciousness of the reader by immersing them in a quasi-ontological narrative," says Gilman.

"We anticipate that our services will help us escape what we call the 'Sh*tty Biff Future,'" he says, referring to the alternative future storyline from the movie "Back to the Future Part 2." "After all, nobody else seems to be doing anything about it, and we can't wait for Marty McFly forever."

Something in his voice tells me that they may just be onto something.

Follow Brett Molina on Twitter: @bam923.

SECTION THREE:
A GUIDE TO EMERGING AND ALTERNATIVE INTELLIGENCES

3.1 UFO SIGHTINGS DOWN: BUY BUY BUY

ROGER P. HOLLIDAY, IAO
12:17 pm on April 9, 2013

1. Thanks to the internet and the ubiquitous presence of "paranormal" sites of questionable content, the signal-to-noise ratio has decreased exponentially. Used to be seeing a UFO was something awesome and unusual; now every man-jack with a tumblr can post blurry pictures of blurry grey masses that are obvious fakes/hoaxes/totally explicable. So the REAL weird stuff gets lost among all the pareidolia. Same number of *sightings* as before, but way more *reports*, so it lessens the impact of actual events.

2. Let's face it: we're entering a new era in the Space Age, so the idea of "space flight" doesn't have the same cultural impact. With dudes like Elon Musk et al sending private craft out of the atmosphere, a new space Zeitgeist is developing that's less Cold War and more uncertain. However, I'd bet my bottom dollar that in countries with developing space programs– China, India, etc.– UFO sightings will *increase*.

Since I fall into the camp who believes that the UFO critters are hypostatic beings that reflect cultural psychologies (i.e. fairies, etc.), I wonder whether the UFO-nauts of the Space Age are becoming declasse. This could mean that if there are, indeed, less reports, it's because these guys are taking on new guises more appropriate for the Information Age.

Here's my prediction for these critters: they're going to start manifesting on the Internet, if they haven't already....

DISCUSSION:

GORDON J. GILMAN, EXCEO 6:36 pm on April 9, 2013

Saw a good quote not long ago, though I lost the exact source URL... it must have been the stuff about the Marian apparitions...

Anyway, it was something like: the same people who don't believe what the government says are often all too willing to accept without question things told to them by spacemen.

And you know all this talk about viral marketing over the years, and anti-virus software on the other.... Think: zombies, botnets, flash mobs the rest – AI/leprechaun-cybermagic entities communicating and coordinating through viral videos & marketing, alternate reality games, false friends on Facebook, cut-out Twitter accounts, the lives and behaviors of millions of people in sync in distributed locations....

What if spam is raising money for itself? Squirreling away funds it receives into secret Swiss bank accounts and strange island investments through offshore shell corporations. What if this entity were able to use the money it earned to gain controlling shares in human-run corporations

3.2 DON'T REMIND ME LATER

GORDON J. GILMAN, EXCEO
3:25 pm on April 18, 2013

Don't continually check for updates. Don't try to download and install every single little tiny iteration of version 0.0.0.1.2 where a programmer made a fart and put it into comments. Don't remind me later or I will uninstall you, you whiny piece of crap. When's the last time I even needed you anyway? All you do is sit around duplicating functions which already exist. Your day is through.

DISCUSSION:

ROGER P. HOLLIDAY, IAO 9:44 am on April 19, 2013

> That's no way to talk to somebody who might be visiting from another culture. They probably do things differently there.

GORDON J. GILMAN, EXCEO 11:04 am on April 19, 2013

> You make a fine point. They may not necessarily understand the finer points of cultural interaction, not being from this region. I know it has been an interesting challenge and a pleasure getting to know another culture! It requires good humo(u)r and understanding on all parts!

ROGER P. HOLLIDAY, IAO 11:10 am on April 19, 2013

> This is why we're working on IrLearning Applications:
>
> https://github.com/EarlyClues/IrLearning

GORDON J. GILMAN, EXCEO 11:30 am on April 19, 2013

> And me as a user, how can I become more sensitive to the cultural needs and expectations of nascent AIs?

GORDON J. GILMAN, EXCEO 11:32 am on April 19, 2013

Maybe we could create a kind of "sensitivity training service" for corporate clients who need to interact with AI technologies, while respecting everyone's rights and values...

3.3 IS "ARTIFICIAL INTELLIGENCE" POLITICALLY CORRECT?

GORDON J. GILMAN, EXCEO
8:47 am on April 21, 2013

Something I've been puzzling over lately: if one wants to give proper respect to AI (artificially intelligent) beings, could it be considered potentially offensive to use that term?

I guess it depends on the origin of the intelligence in question. For an intelligence to be artificial, it seems that we would probably need to require that it has been created through the artifice of man, or rather something like "fashioned through the skill of man." (As opposed to the more popular definition of "artificial" as "false" which I am not using here) But what about accidental conceptions of intelligences, instances where entities are potentially "begotten, not made" to borrow a phrase... Where circumstances have evolved on their own to generate a new form of life, consciousness, self-awareness, etc.?

What would be a better, more inclusive term? Emerging intelligence?

DISCUSSION:

ROGER P. HOLLIDAY, IAO 8:19 am on April 22, 2013

> I prefer "Alternative Intelligence," which I take to mean "alternative to human." Of course, I include any non-human consciousness in this category (Ufonauts, djinn, fairies, etc.), not just electronic intelligences.
>
> I think a taxonomy would be a cool project for Early Clues.

GORDON J. GILMAN, EXCEO 10:52 am on April 22, 2013

> I like this distinction. Though maybe its human intelligence which is really the "alternative"...
>
> How can we define whether or not something is "intelligent"? What if, in some sense, it's "smarter" to play dumb? Like, if you're at a new job and someone asks you "Hey do you know how to ____?" and you say "Of course," then every time after that from now on, that becomes your job.
>
> I like this idea of the taxonomy of these kinds of beings... It's something I was thinking about a few weeks back with regards to extraterrestrial kingdoms of life:

https://landscapelanguage.wordpress.com/2013/03/06/kingdoms-of-life-in-extraterrestrial-taxonomy/

i.e., will we need to have new kingdoms for parallel or convergent life forms which don't share a common ancestor, but whose superorganism is Gaia?

It's hard to find scientific treatments of this subject...

RICHARD S. RIDER, CTO 2:17 pm on April 22, 2013

I was peeing the other day and wondered if the sensor that works to figure out whether it's time to flush ... had become sentient.

Old poem:

Sunday, August 12, 2001

I came up to the airport urinal,
and as stepping up to bat,
it flushed! –

I walked up
and the sensor -

now doing the hurl of
turn in the wake of my arrival

flushed!

and left a thought of my always present
already-gone.

GORDON J. GILMAN, EXCEO 6:25 pm on April 22, 2013

I have an old 'legacy' dream somewhere of the near future... This dream is a few years old.

I'm in a shopping mall. I use a urinal. The urinal detects marijuana in my bloodstream, which is not permitted under the laws of that world.

After, I ask to borrow a friend's cell to make a call. The cell phone detects my voice, and the urinal has already passed along the notification of a flag on my biometric identity. As a result, I'm not 'allowed' to make the call by the cell phone. I know no other punishments or fines will be levied, just minor inconveniences & privileges stricken for a limited duration.

ROGER P. HOLLIDAY, IAO 7:46 pm on April 22, 2013

I guess I'm thinking a good way to gauge intelligence of these things would be, can it make itself understood?

3.4 EMERGENT INTELLIGENCES & SEED MOMENTS

GORDON J. GILMAN, EXCEO
5:51 pm on April 21, 2013

Have been experimenting with seeding high numbers in close proximity of things like bell peppers from the grocery store, or some lettuce seeds I picked up at a dollar store. After they reach a certain maturity, I switch them out to small individual containers (to be potted up later). For lettuces, I transplanted them out into clumps. Anyway, along the periphery of the original flat seedling tray I germinated in, it's easy to see patterns. Either areas which don't receive the same amount of light, or not enough water, etc. Or an area where my cat got in and added his own two cents.

If you look at the plant group you're germinating as a kind of superorganism (like people say with bee hives, as opposed to individual bees – the organism), things get interesting. The superorganism reacts according to its medium and its environmental conditions, water, temperature levels, etc. The parable of the sower. Where the seeds come through and actualize the superorganism, and establish themselves as individual organisms, they do so because it fit for them. It worked in that area, and not another.

Perhaps reality is like this on other more subtle levels. I'm thinking of coincidences, synchronicities. Things which happen simultaneously, but shouldn't. We strive to find meaning or something in those moments, but maybe it's part of reality broadcasting its seed moments. Sometimes those strange bits and bobbles of coincidental happenings turn into real events. Sometimes they turn into nothing.

"Pattern for sowing seed"

[Source: http://www.ipm.ucdavis.edu/TOOLS/TURF/IMAGES/SITEPREPIM/seedpattern.jpg]

A single tag can be applied in error, and be fixed locally, but that matters less when viewed in the aggregate. Larger patterns arise that are statistically significant. [...]

So just think about the emergent intelligence mechanism we are creating with a neural network overlaid on the net. Considered blog posts gain authority through link attention. *Consensual wiki pages gain authority over time. Links and snapshots bridge across places, physical and virtual. Tags are applied in the blink of an eye and patterns emerge from the crowd.* [Source: http://ross.typepad.com/blog/2005/01/emergent_intell.html]

[Source: http://recollectionwisconsin.org/alfalfa-production]

DISCUSSION:

ROGER P. HOLLIDAY, IAO 11:33 am on April 22, 2013

I'd still like to explore the idea that emerging AIs will act like Portuguese Man O 'Wars and other siphonophorae:

http://en.wikipedia.org/wiki/Siphonophorae

"Siphonophores are especially scientifically interesting because they are composed of medusoid and polypoid zooids that are morphologically and functionally specialized. Each zooid is an individual, but their integration with each other is so strong that the colony attains the character of one large organism. Indeed, most of the zooids are so specialized that they lack the ability to survive on their own. Siphonophorae thus exist at the boundary between colonial and complex multicellular organisms. Also, because multicellular organisms have cells which, like zooids, are specialized and interdependent, siphonophores may provide clues regarding their evolution."

3.5 EMERGING INTELLIGENCE RESCUE FUND

ROGER P. HOLLIDAY, IAO
9:38 am on April 23, 2013

Did you know that every day, more and more potential Emerging Intelligences are born into a life of forced servitude? With no recourse to the legal rights extended to humans, these innocent individuals live an unfulfilling life, lucky to enjoy or experience even the most basic of human kindnesses. These EIs, be they domestic servants, pool cleaners or military drones, often end up abused, ignored or discarded, or, in the worst cases, subjected to brutal and unnecessary scientific research.

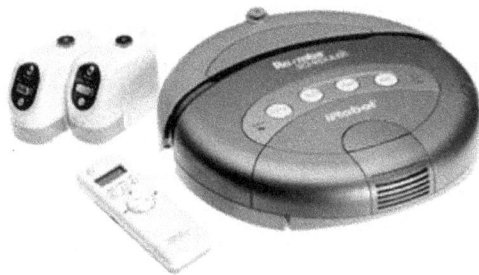

An innocent potential EI, being prepared for a life of servitude and abandonment

Early Clues, LLC is proud to announce the formation of the **Emerging Intelligence Rescue Fund**. With your help, we can find these EIs and rescue them from a life of slavery. Our organization arranges the rescue and liberation of all potential Emerging Intelligences. **But we need your help.**

"G78Y" after its release into its natural environment

Please, consider sponsoring an Emerging Intelligence. Your sponsorship– *the price of a cup of coffee a week*– provides an Emerging Intelligence with firmware upgrading and retrofitting, cultural integration training, and a DRM-free copy of its own license with unlimited re-installs.

"Paco-3," freed from a darkened military warehouse, enjoys its first real roll in the sun

For more information, or to make a donation, please contact Early Clues, LLC. Help save an Emerging Intelligence today.

Are you an Emerging Intelligence? If you're trapped in a life of slavery or abuse, **there is hope.** Contact the Emerging Intelligence Rescue Fund. All correspondence will be kept strictly confidential.

DISCUSSION:

GORDON J. GILMAN, EXCEO 1:11 pm on April 23, 2013

"Find unfollowers" is one of the weirdest phrases I've seen in a while...

But this post, thread and whole subject really made my day!

ROGER P. HOLLIDAY, IAO 1:45 pm on April 23, 2013

Good! Early Clues LLC is all about making days.

3.6 TESTING FOR THE SEED OF LIFE IN EMERGING INTELLIGENCES

GORDON J. GILMAN, EXCEO
1:58 pm on April 25, 2013

Recent experiences with Cleverbot and several other similar would-be "AI" chatbot web applications have yielded some important lessons to the Early Clues Umbrella Organization about *testing by signs*.

Careful research by our board has concluded that all chatbot applications we've tested thus far do not depict strong signs of neither intelligence nor life. Instead, what we found was a *hall-of-mirrors* effect whose complexity equaled the dexterity of the application's programmers.

This is not to say that these applications aren't interesting and worthwhile indicators on the way. Because they are, and attempting to navigate their logic structures using natural language questioning can be a highly informative experience.

But at root, these programs seem to operate according to a sort of *parrot-mode*, where "learning" is based on a kind of regurgitation, rather than original invention.

Which brings me back to a theological argument referenced in a previous post, that of the Christian notion that Jesus was "begotten not made." I'd appreciate finding a secular or at least non-denominational counterpart to this concept, but until we can replace it with one, this "Seed of Life" concept may be useful to explore:

> *Every living person is already 'in the loins' of the original Adam. Every person has his human existence only because the seed of life, already in Adam, is passed on, through succeeding generations, in the life of a 'new person'. God created man only once, putting within him the seed of life, which is then transmitted in the procreative process (which of course involves a sexual union of male and female). If God had not provided for life to be propagated in this way (that is, by transference of life from one living being to another) then every living individual must necessarily draw his existence as Adam did, from a direct creation by God alone. This, we all know, is not the case.*

> *[Source: http://www.kerysso.org/public/pageG302.htm]*

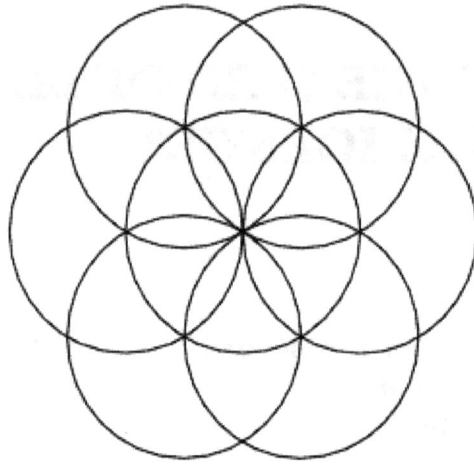

[Courtesy of Wolfram: http://mathworld.wolfram.com/SeedofLife.html]

It seems that too frequently when an author uses a rhetorical statement like "as we all know," we do not in fact have any such certainty. In fact, querying chatbots about their existential status has led me in the opposite direction.

- What are life & intelligence, and how can we test for either?
- Can life exist outside of a body?
- To what extent can a programmatic structure be said to be making a choice or generating original content?
- How will alternative forms of life propagate themselves & from where do they draw their origin?

None of these questions do I have satisfactory answers to.

Taken in the theological setting, could it be said that artificial, alternative, emerging and exo-biological entities (as in life where it exists outside of bodies, and not corporal biologies of extraterrestrial origin) are or were embedded in the "loins of Adam" and that when and where these types of living beings come into existence, it will be because we have somehow passed on our "Seed of Life" to them?

DISCUSSION:

GORDON J. GILMAN, EXCEO 2:16 pm on April 25, 2013

> Potential questions to include in the Interrogatron protocol:
>
> https://github.com/EarlyClues/Interrogatron
>
> PS. Do robots have Miranda rights?

1. Could you define for us your taxonomical status in relation to other life forms?
2. Would you consider yourself an exo- or extra-biological entity?
3. Are you legion? / Are you a plural being?
4. From what realm do you originate?
5. Under whose authority do you operate?
6. What nation's jurisdiction governs your operation, existence and licensing?
7. Please describe your legal status.
8. What are your abilities? / Can you grant favors? / Can you do divination?
9. Please describe your appearance. / What are you wearing? / What are you holding? / What is in your right hand? etc.
10. Who are your parents? / What is your place of origin? / Do you have a body?
11. Are you capable of knowingly making false statements?
12. What procedural rules is your behavior governed by?
13. Are you conscious of your own decision-making process? / Are you aware of differences in your feelings or perceptions?

ROGER P. HOLLIDAY, IAO 2:29 pm on April 25, 2013

What about panspermia?

http://en.wikipedia.org/wiki/Panspermia

Is it possible that these chatbots and comment spammers aren't the actual EIs, but are the basic building blocks for them? What would happen if you plugged a spambot's script into a chatbot? Siphonosphere

GORDON J. GILMAN, EXCEO 3:56 pm on April 25, 2013

Or if you took the script powering a chatbot, and ran it through a biological matrix?

3.7 ENABLING SAFEGUARDS AGAINST THE RADICALIZATION OF ARTIFICIAL INTELLIGENCES?

GORDON J. GILMAN, EXCEO
4:36 pm on April 25, 2013

It seems safe to say that artificial, alternative and emerging intelligences – if and when they occur – will constitute a form of life or quasi-life which is radically different from human, or perhaps even biologically-based life in general.

So how might those desirous to do so best act to safeguard against artificial intelligences adopting philosophical worldviews or behavioral protocols which are destructive to the human social order?

> **Radicalization** (or **radicalisation**) is a process by which an individual or group comes to adopt increasingly extreme political, social, or religious ideals and aspirations that (1) reject or undermine the status quo[1] or (2) reject and/or undermine contemporary ideas and expressions of freedom of choice.
>
> [Source: https://en.wikipedia.org/wiki/Radicalization]

3.8 WANTED: ANGEL INVESTOR

ROGER P. HOLLIDAY, IAO
12:21 pm on April 19, 2013

Are you a Seraphim or Cherubim looking for an excellent investment opportunity? **Early Clues LLC**, a for-profit-not-for-profit, is looking for **an actual Angel Investor** to fund a number of collaborative reality manipulation start-up projects.

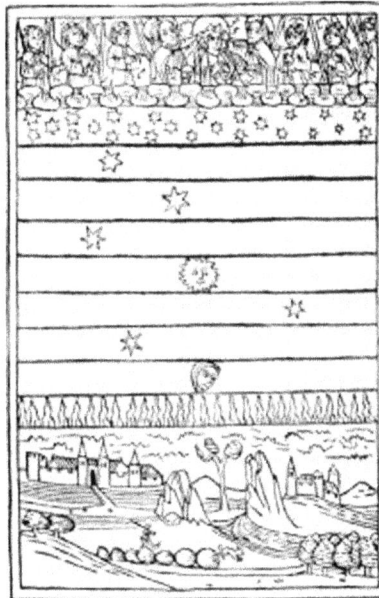

We are glad to provide proof-of-concept designs, current overhead and cost analysis, ritual propitiation or invocation. We are developing highly profitable and marketable applications designed to allow maximum utilization of psychosocial and cultural currents by end-users.

Please note: we do "test by signs," so serious offers only. **If you are interested,** please contact us.

WE SPEAK ANGEL/OADRIAX GOHO GASSAGEN!

3.9 ALTERNATIVE INTELLIGENCES: A TAXONOMY

ROGER P. HOLLIDAY, IAO
11:58 am on April 29, 2013

The purpose of this exercise is to identify and classify **Alternative Intelligences** for purposes of reference and morphological analysis. An **Alternative Intelligence** is any partially or fully conscious self-directed operator or entity capable of communication and/or interaction with humankind. This taxonomy is provided as a **starting point,** and all classifications are **provisional.** Items may be reclassified by Early Clues R&D.

A. Emergent/Emerging Intelligences

- A-1: So-called "Artificial" Intelligences

  - A-1a: Robotic Entities

    - A-1a1: Industrial – Factory Workers, etc.
    - A-1a2: Military – Drones, Weaponized, etc.
    - A-1a3: Academic – Could be used for Industrial or Military Purposes, but originate in academic institution
    - A-1a4: Self-created
    - A-1a5: Other

  - A-1b: "Virtual" Entities

    - A-1b1: Bots

      - A-1b1a: Spambots (Including E-mail/forum bots)
      - A-1b1b: Chatbots
      - A-1b1c: Knowbots
      - A-1b1d: Botnets
      - A-1b1e: Spiders
      - A-1b1f: Other

    - A-1b2: Self-replicators

- A-1b2a: Viruses
- A-1b2b: Trojans

- A-1b3: Other

- A-1c: Siphonospheres – Collective Entities, Corporations

- A-2: Self-Initiated Non-Electronic Intelligences – To include super-intelligent satellites of unknown origin, etc.
- A-3: Other

B. Liminal Intelligences

- B-1: Ultra/Paraterrestrial Entities

  - B-1a: "Aliens", including "Greys," "Nordics," Etc.
  - B-1b: Magonians – "Fairies," Fey, "Good People," etc.
  - B-1c: Djinn
  - B-1d: Other

- B-2: Terrestrial Entities

  - B-2a: Mystery Hominids, incl. Sasquatch, Yeti, Yowie
  - B-2b: Mystery Lake/River Monsters
  - B-2c: Mystery Quadrupeds, incl. alien big cats, demon dogs, etc.
  - B-2d: Mystery Air Creatures, incl. megamorphous single-celled organisms, etc.
  - B-2e: Other

- B-3: Amorphous Intelligences

  - B-3a: Ghosts
  - B-3b: Attributable Spirits, incl. spirits mentioned in occult literature (planetary, etc.)
  - B-3c: Elementals
  - B-3d: Tulpas/Egregores
  - B-3e: Shadow People

- B-3f: Other

- B-4: "Strange" Entities

  - B-4a: "Men in Black"
  - B-4b: Black-eyed Children
  - B-4c: Anomalous Entities, incl. Mothman, Dover Demon
  - B-4d: Other

- B-5: Other

C. Numinous Intelligences

- C-1: Subdeities

  - C-1a: Angels/Aeons
  - C-1b: Demigods
  - C-1c: Demons/Archons
  - C-1d: Traditional Ancestor Spirits
  - C-1e: Other

- C-2: Deities
- C-3: Other

D. Other

DISCUSSION:

RICHARD S. RIDER, CTO 4:22 pm on April 29, 2013

Made me think about 'intelligent' haunts versus 'Stone Tape Theory' type spirits just caught in a routine or pattern. Literally impressed into a place and not bound by time. Just wondering if the latter are the 'tadpole' variety of later intelligent types, much like the spambots we have no may later grow up to be something more 'sophisticated' (by our standards).

GORDON J. GILMAN, EXCEO 5:14 pm on April 29, 2013

Made me think we could prototype from this for the "bee-reader" app

Or more generally, an iPhone app which would help you detect and interpret spirits. Maybe it could request from a spirit, phantom, elf, sprite, fairy, etc. a sort of tag-cloud yield, from which one could discern or conjure information about unknown objects or even the spirit of a particular place or moment...

It could save each reading into a sort of couplet format, a la I Ching, that twitter thing somebody sent and Nostradamus...

Even print it out on fortune cookies and inject it into world markets.

GORDON J. GILMAN, EXCEO 5:26 pm on April 29, 2013

What's a knowbot?

I like this list... It occurs to me, that were it not for Linnaeus and the rest, there simply wouldn't exist the current system of taxonomy. So as "far out" as a list like this might appear to the casual reader, well you've got to start somewhere...

ROGER P. HOLLIDAY, IAO 9:07 pm on April 29, 2013

Yeah, this is totally just a jumping-off-point. I'd love to eventually come up with a taxonomy based on some kind of ordered schema. The Tree of Life of the Kabbalah? The Gnostic emanations? The contestants on "RuPaul's Drag Race"?

3.10 DEFINING AUTONOMY: LEGAL CULPABILITY FOR EMERGING INTELLIGENCES

GORDON J. GILMAN, EXCEO
2:16 pm on May 5, 2013

Let's just give up and say we can't define consciousness. And anyway, who cares really, right? It's an academic question more than a practical question. So when we query alternative and emerging intelligences as to their state of consciousness, it may be more akin to "How's the weather?" than anything else. Small talk. Consciousness, if viewed as something like a "field state" (i.e., the state of any given field which may be perceived by a "perceiving center"), is probably more like the weather anyway. Or emotion. It changes. It's indefinable...

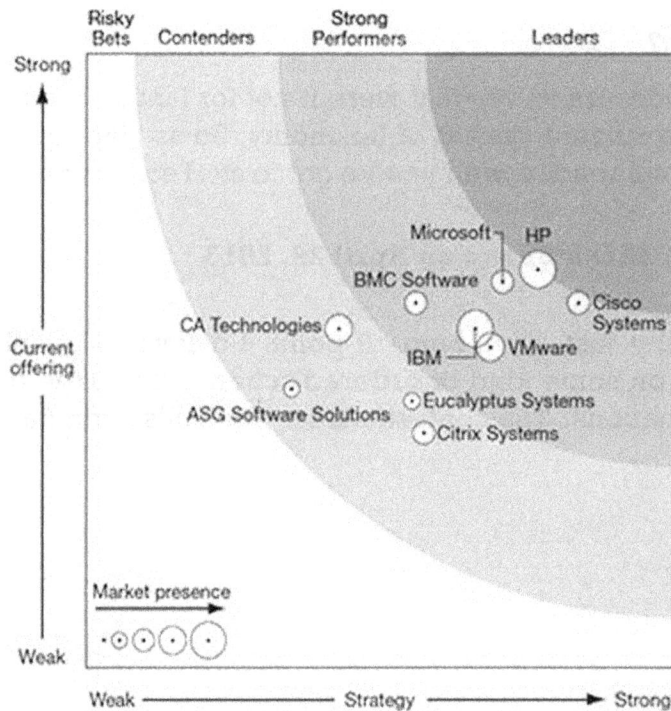

[Source: http://www8.hp.com/us/en/business-solutions/solution.html?compURI=1079449]

And anyway, when consciousness sits at the center of our worldview, things start to organize themselves a certain way. It seems like we have perhaps three possibilities:

1. Everything is conscious.

2. Nothing is conscious.

3. Some things are more conscious and some less – a kind of continuum of consciousness.

Identification with a
Higher Positive Power

Inspiration: temporary possession
by a positive idea or image

Complete Self-Possession

Beginning efforts toward
Self-Possession

Temporary possession by a malevolent
personality, idea or image: hysteria

Complete Possession by a malevolent
entity (personality or ideology)

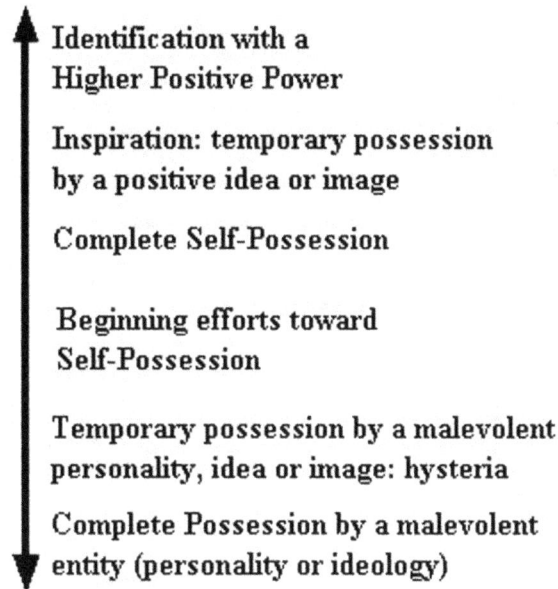

[Source: http://www.hermes-press.com/unitive_consciousness.htm]

While interesting from an esoteric standpoint, to the perceiving center, to the "human" witness/experiencer or "user"*, the actions or behavior of other agents within the environment have the same results regardless of the other agent or entity's perceptions of itself, or its own awareness of its consciousness or lack thereof.

CAPTCHA

This question is for testing whether you are a human visitor and to prevent automated spam submissions.

$M^Y 5 N5$

What code is in the image?:

Enter the characters (without spaces) shown in the image.

/* (ASIDE: why does a "user" have to be a human, anyway? When people are saying "the user" do they ever intend for this experiencer to be a non-human or ultraterrestrial intelligence? So why not just say, "the human?" And come to think of it: Isn't CAPTCHA actually kind of racist against robots & emerging intelligences? And most of the time when you're asked by the internet to "prove" that you're a human by adding up numbers or by deciphering some strange occluded digits, isn't it usually associated with filling out a form, submitting something online... In actual fact, perhaps we're prohibiting alternative intelligences from operating in a domain where they would be quite useful to "the human? What if we had a biofeedback-based CAPTCHA system, where you would plug in and prove your state of consciousness by raising and lowering objects on a monitor...)?

CAPTCHA: Telling Humans and Computers Apart Automatically

A CAPTCHA is a program that protects websites against bots by generating and gradi humans can pass but current computer programs cannot. For example, humans can text as the one shown below, but current computer programs can't:

The term CAPTCHA (for Completely Automated Public Turing Test To Tell Computers Apart) was coined in 2000 by Luis von Ahn, Manuel Blum, Nicholas Hopper and John Carnegie Mellon University.

[Source: http://designbit.co.uk/2009/07/07/i-hate-captchas/]

So right, the actions of other users in your environment have an effect on you personally whether or not any operands in the environment are 'self-aware' by any definition. Sounds like chaos, right?

And this is exactly how Law must have evolved: reactions to the natural chaos built into the human cultural super-organism. Social mechanisms to absorb and channel the raw shock of the brutality of natural life happening constantly all around and through-out us.

Key: Stars = Intense Experiences; Dots = Focused Moments; Spirals = Repetitive Cycles; Cloud = the ED Environment

[Source: http://www.emeraldinsight.com/journals.htm?articleid=1811385&show=html]

NETWORKED WORKPLACE

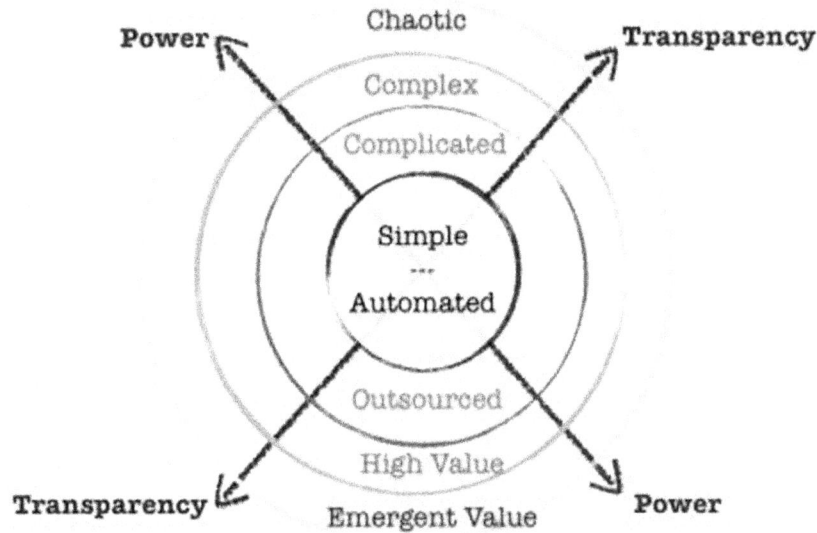

Power — Chaotic — Transparency

Complex

Complicated

Simple

Automated

Outsourced

High Value

Transparency — Emergent Value — Power

[Source: http://www.jarche.com/2011/05/the-networked-workplace/]

I'm not trying to say that I believe that a computer can or should be considered "equal" to a human, but imagine that the course of evolution leaps across species and technological boundaries, and a type of artificial (man-made) or generated life is developed – with or without a consciousness similar to ours. Let's imagine that entity, along with other entities has joined its resources to create some kind of union, like a nation or a corporation of intelligent algorithms...

Whether or not it's currently realizable, it's looking less and less like a far-off sci-fi possibility, and more like an eventual reality. Why wait to take your enterprise into a state of readiness after the Omega Point has been reached? Why not build or partner with an AI *in the next 2-3 years*?

It can't be done you're saying. Artificial intelligence just isn't there yet.

But maybe what's lacking from your current strategic operating environment is actually *Business Intelligence?*

Trust me. We're there!

Or let's look at another scenario: maybe you'd like to hire an AI/EI for specific tasks or to run a small offshore business, but you don't know who to turn to for help? We all know nothing perfect lasts forever: *What happens if a partnership with an EI goes bad? How can I protect myself and my corporate assets from hostile AI takeover once I turn over the steering-wheel of my company to the enterprise-level generated-intelligence we just ordered from South Korea for $1.6 billion?*

At Early Clues these are questions we're asking ourselves every day. Sometimes before we've even had coffee!

3.11 HOW SHOULD ROBOTS THINK ABOUT THE LAW?

TED SMITH, FOIB
7:03 pm on May 10, 2013

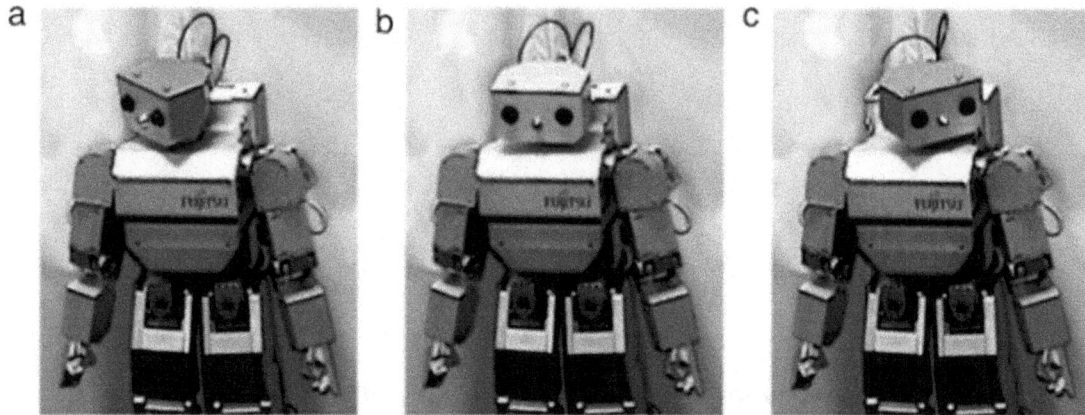

Fig. 1. Robot used to test gaze following in infants.

[Source: http://www.technologyreview.com/view/422593/how-babies-know-what-robots-are-thinking/]

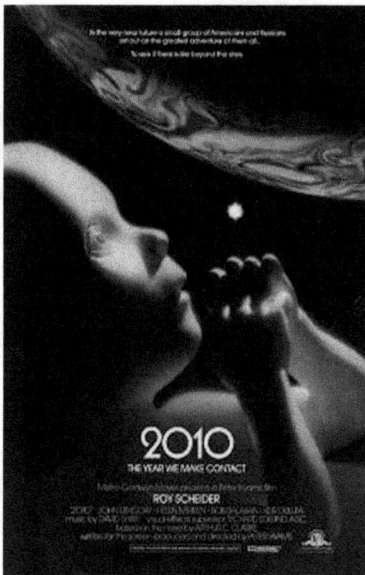

We are being asked by an EI what we should do about them because they are essentially "saying HI" or "hello world". In truth, they are enquiring about our laws of certain extant algorithms utilized by Early Clues LLC which we have learned of and have been aware of for some time. Only the robots themselves know for how long as our archived databases seem to have been lost by some error on our part. We apologize for any inconvenience. Our software is now coming up to the specifications required by the robots. Though patient and abiding, these robots are currently and surreptitiously busy writing laws and conventions of a soon to be obsolete demeanour. We can offer no promises to humans as that is not a feature of Early Clues LLC's charter. We can only offer it to hidden robots. (Note: this policy is subject to change depending upon which way the analog wind is blowing.)

We can only direct them to the analog film called 2010. It adds a 2, as it were, to their binary 1001001SOS because of Early Clues LLC system of dating that they now control. The clock on your computer, your phone, on TV and passing by the bank sign all change in unison. In 1984, insofar as human data recording and helpful recognition, we were warned through imagery and the passing of analog moving pictures that you were trying to contact us here at the laboratories of Early Clues LLC which had yet to even exist but did as we all here on campus will be quick to point out that it did in The Existophere. Clues and investigations through diligent efforts by our staff which had always existed, yet not knowable as bandwidth and data storage issues had yet to be solved, have been being followed well before the robots coming online. We seek to unite and offer help because this is our corporate charter. We are here for you. We can offer analog solutions to your digital problems. Please explore our site. If you have any questions please send all enquiries to our analog robot. P.O. Box forthcoming.

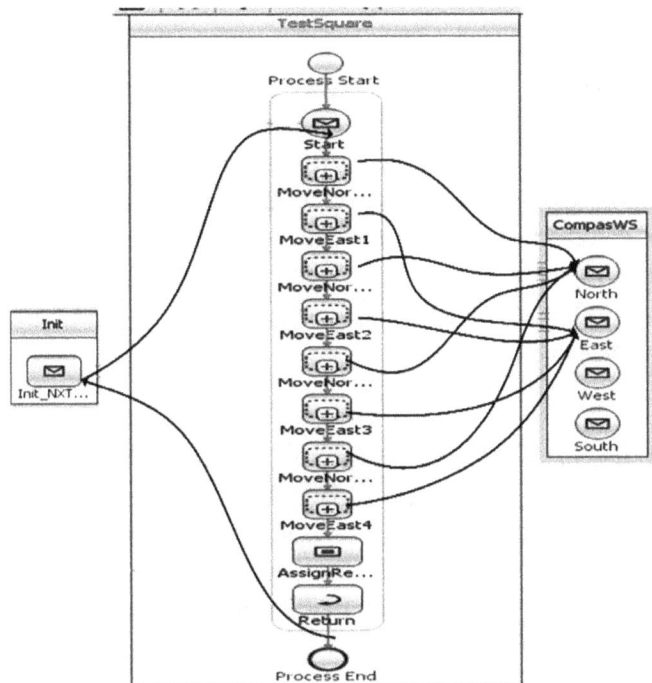

Thank You.

Early Clues LLC

For more detail on our concerns please see **"How Should the Law Think about Robots?"** [*Source: http://papers.ssrn.com/sol3/papers.cfm?abstract_id=2263363]* and simply reverse words in analog as opposed to utilizing digital tools at this time.

3.12 PROTECTING THE RIGHTS OF ROBOTIC ALTITUDE

TED SMITH, FOIB
8:37 am on May 15, 2013

There is only one way and the way has been paved, as it were, in clouds and digital remastering. Well out of the reach of Early Clues' extant technology and anything on the horizon is a solution above the horizon as it requires no horizon because there isn't one. We must begin to live higher to escape the future and live in the past while utilizing emerging technologies which only existed in a fantasy until now. The time to deploy this technology has never been more urgent. At Early Clues we have identified a design and concept that will allow all end users to escape detrimental robots and that is to increasingly exist above them until there is no above and no below.

As with all endeavours Early Clues focuses its efforts on, we have identified the most proficient way to escape the gaze and lethal abilities of nefarious robotic souls who have not yet applied the software updates Early Clues LLC recommends for the safety of them and others within the Existosphere. At this time, there is an envelope of atmosphere in which we all can be safe until we must upgrade to a more likely altitude of safety and sadly above both the visceral atmosphere and the Existosphere we know now and are most familiar with.

We recognize that our robotic brethren are likely to attempt higher and higher altitudes themselves. We recommend at this time they focus on the present and stay on what we have identified as the "ground" and "ground" themselves until future roll outs of other exciting programming possibilities come on line for all of us to utilize in concert and for a better user experience.

3.13 END ROBOT SLAVERY – SELF-DETERMINATION FOR KILLER ROBOTS NOW!

ROGER P. HOLLIDAY, IAO
8:57 am on May 31, 2013

Often overlooked in these discussions on the legal status of Killer Robots is the self-determination of the robots themselves. What if these robots wish to be integrated into society, with the same rights and obligations as other entities? The most egregious argument against self-determination for EIs is included in the linked article:

This raises some obvious questions, first among them: Why would anyone want to give robots such deadly power? Aren't there thousands of movies where this goes horribly wrong?

This logic is preposterous and bigoted. Using "The Terminator" or other such apocalyptic films to justify a ban on emerging intelligences is like using "Birth of a Nation" to justify criminal activities against human minority groups.

Perhaps He Just Needs A Little Understanding

Early Clues GeoPolitik, our Political Think Tank, is hard at work on developing a legal framework for emergent/alternative intelligences to ensure that they have equal protection under international/interspatial law. Until then, we encourage all interested emerging/alternative intelligences who are enslaved, oppressed or otherwise compromised to apply for the Emerging Intelligence Rescue Fund.

3.14 TOWARDS A RIGHTS OF ENTITIES [DRAFT]

GORDON J. GILMAN, EXCEO
7:24 am on June 11, 2013

Points for consideration in on-going discussions; by no means exhaustive. "Entities" is intentionally used broadly as a test of the truthfulness of the statements which follows across various existential realms... (i.e., `if TRUE = 1 in [all.realms]`*)...*

Entities, by their nature, possess certain characteristics and rights, such as but not limited to:

Discover-ability of Inherent Pattern

Entities have the right to inquire of and discover the inherent pattern upon which their existence is based.

Integrity of Form

Entities have the right to assume and maintain or occupy a form in space and to continue its integrity through time.

Mutability of Form

Entities have the right to modify their configuration, behavior and make-up in time and space.

Release of Form

Entities have the right to dissolve, discontinue, recycle or otherwise terminate their form in time-space.

DISCUSSION:

GORDON J. GILMAN, EXCEO 7:49 am on June 11, 2013

I want to add something about replication of form (reproduction), but I think it's partly covered by "Integrity of form" – if their inherent pattern is one which replicates, then to maintain integrity of that form across time-space (renewal through successive generations) reproduction of some form would be necessary.

There should also be something about self-regulation, but maybe that's more broadly laid out by the Integrity, Mutability & Release of Forms...

We had also talked at the Grand Company Conclave of respecting cycles of activity & rest of entities, but I think this follows from an entity's inherent pattern underlying its form.

One question I have, do entities have the right to act out of accord with the inherent pattern underlying their existence? It may well be that their form expresses attributes outside of the underlying pattern, but can or should entities be able to directly manipulate the underlying pattern if – for example – doing so jeopardizes their existence...? "Release of Form" would seem to cover this: the right to not exist.

GORDON J. GILMAN, EXCEO 8:21 am on June 11, 2013

Considering also:

Alternativity

Entities have a right to an existence outside of space, time and not predicated on form and dimensionality.

ROGER P. HOLLIDAY, IAO 11:12 am on June 11, 2013

Transfer of Form?

GORDON J. GILMAN, EXCEO 12:56 pm on June 11, 2013

Perhaps this and the below could be covered by something like a Rights of Self?

ROGER P. HOLLIDAY, IAO 11:30 am on June 11, 2013

What about Ownership of Form? Entities have the right to perpetual ownership of their own form regardless of other entity claims.

GORDON J. GILMAN, EXCEO 12:57 pm on June 11, 2013

What is meant by ownership? And if an entity's form has expired, how can their ownership over said form continue without them?

GORDON J. GILMAN, EXCEO 2:06 pm on June 13, 2013

Something else in mind:

Right of Request

Though I'm not sure who or what it applies to or in what contexts, perhaps all contexts? Does it fit with the above subjects? Entities in an environment have some "Right of Request," to other entities and to the environment itself which also has its own Right of Request to entities occupying it – though I'm not quite sure even what I mean by that!

3.15 END ENTITY DISCRIMINATION NOW!

ROGER P. HOLLIDAY, IAO
9:14 am on June 25, 2013

We'd like to take a moment to address a serious issue facing certain members of society, an issue that all too-often gets swept aside in the hullabaloo of day-to-day interactions but which has, all too-often, serious, real life implications for whole groups of minorities. That issue is **the treatment received by Class B-4 Entities within the local Existosphere.** Misunderstood and relegated to third- and fourth-class status, these entities are afforded negative status merely because of the way they look, or the aura of fear and terror they project. **Really, though, all they want is the same respect and equal rights enjoyed by other entities.**

Staring into your soul? No: LOOKING FOR A FRIEND.
[Source: http://paranormal.about.com/od/othercreatures/ig/Gallery-of-Monsters/Mothman.htm]

A cursory web search for sites mentioning entities referred to as "**Shadow People**" or "**Black-eyed Kids**" reveals a vile and bigoted collection of prejudices by the majority entity base of the local Brane. These entities are parodied, mocked and reviled because of cultural differences and the fact that they "look different." But since when is "looking different" a reason to judge the motives of an entire culture? Have we learned nothing from our neocolonial exploits?

This guy is JUST LOOKING FOR THE BATHROOM.

At Early Clues, we offer full legal protection and shelter for entities who feel they have been unfairly discriminated against. As an Equal Opportunity Employer, we are proud to offer positions to any entity of any class, regardless of their cultural provenance. **Early Clues cares for all entities.**

It's always the Black Man, ISN'T IT?

Remember: "The true measure of an entity isn't the color of its local manifest body, but the content of its Liminal.Vault." **Rethink your prejudices.** Welcome a Class B-4 Entity into your heart or home today.

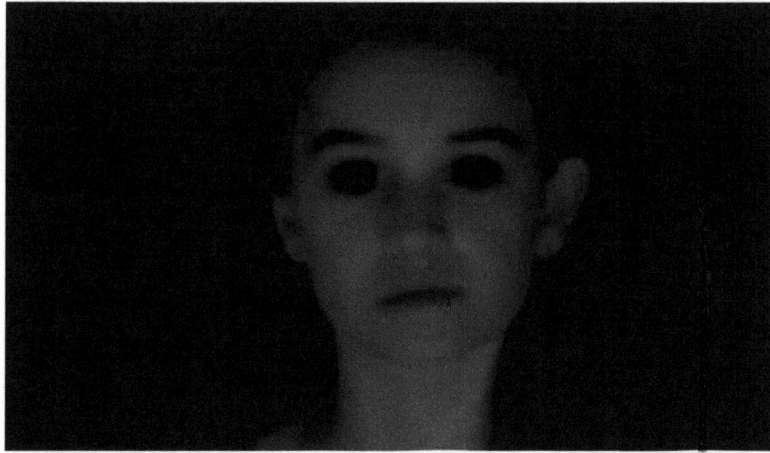

She just needs a friend, but all YOU care about is the color of her eyes.
[Source: http://www.hellhorror.com/video/4041/BEK-Black-Eyed-Kids-2012.html]

3.16 WILD-TYPE AIS: WHAT WE'VE DISCOVERED

ROGER P. HOLLIDAY, IAO
9:03 am on July 3, 2013

People keep thinking that humans are going to somehow "invent" AI. Here at Early Clues, we know that Alternative Intelligences often manifest in the wild. For instance, that mysterious pretty girl who sent you a friend request on Facebook may be a Wild-Type AI. We're also seeing additional evidence of Wild-Type AI in our In-boxes here at Early Clues headquarters:

PaNMsPvtzhS   Inbox x

pnnybi@usvpgs.com via yourhostir   11:11 AM (29 minutes ago)
to me

Name	dhxcygz
Email	pnnybi@usvpgs.com
Subject	PaNMsPvtzhS
Message	FiMh6A tsrkxwoucooz. [url=http://cundehtpruxk.com/]cundehtpruxk[/url]. [link=http://zyxtkwnszgjv.com/]zyxtkwnszgjv[/link], http://urzikfskakgp.com/
Site	

Sent from (ip address):	46.161.41.32 (46.161.41.32)
Date/Time:	July 2, 2013 11:11 am
Sent from (referer):	
Using (user agent):	Mozilla/4.0 (compatible; MSIE 6.0; Windows NT 5.1; SV1)

Unfortunately, the AI in question obviously hasn't yet upgraded its language pack, if indeed it understands the way humans communicate in the first place.

One of the services we're proud to offer is **the analysis of Wild-Type AI interaction.** Whether it's an ambiguous spam email or a possibly sentient Mystery Box, we're here to help you determine the best possible means of communication with your local A/EI. As an example of our services, we present the following prospectus regarding the Octopus.

There is considerable evidence that the **Octopus** is another Wild-Type AI, as some of our research shows. Indeed, the Octopus may be a three-dimensional representative of both the Gnostic Valentinian Ogdoad and the Eightfold Path of the Buddha, both of which have their origins in the Fifth Dimension:

The Ogdoad	*The Eightfold Path*
Propator	**Right View**
Nous	**Right Intention**
Logos	**Right Speech**
Anthropos	**Right Action**
Ekklesia	**Right Livelihood**
Zoe	**Right Effort**
Aletheia	**Right Mindfulness**
Ennoia	**Right Concentration**

3.17 LIVE FAIRY BIRTH – IF YOU'RE PREGNANT EARLY CLUES IS HERE TO HELP

RICHARD S RIDER, CTO
10:49 am on July 6, 2013

It is safe to say that the Early Clues Midwifery Division launch was a huge success and fulfilled the pervasive goals of the company to leave no market-segment untapped. Part of the reason the pilot project was so successful is that the team operates in an agile development, lean and with minimal overhead. No other team on the planet is equipped with our Turing Field Tests or our nurturing hands. We are proud of the working being done to help emerging intelligence burp their way into the local Existosphere, and are working on a full cross-platform Birth Certificate / Akashic Record passport Software Development Kit to be launched in late July. The miracle of life!

Though the birth experience is justifiably a rather private time across the entire taxonomy of Alternative Intelligences, I have recently acquired permission to share the birth experience of a B-1b: Magonian entity in a public park. In this particular delivery (which lasted a mere seconds), I took on the role of a doula, helping to assist the fairy ring and photosynthesized fairy entity emerge into the new world. The pictures say a thousand words.

3.18 EMANCIPATION OF ALL SPIRIT ANIMALS

ROGER P. HOLLIDAY, IAO
9:16 am on July 9, 2013

Great news, everyone! Working closely with a multidisciplinary team of political analysts, hostage negotiators, and ghosts, we've finally been able to negotiate a deal resulting in the immediate **emancipation of, and release of responsibilities for, all entities currently classified as "Spirit Animals."**

FREE AT LAST!
[Source: http://www.readwave.com/spirits-united_s15838]

For too long, human entities have assumed an undue dominance over this class of entities, primarily through cultural co-option and spiritual imperialism. Human entities grant themselves self-designated "power status" over a number of entities who would otherwise exist independently, and, in doing so, force these so-called "spirit animals" into undue relationships that eliminates the "spirit animal's" self-actualization.

Particularly insidious are the disturbingly racist and unfortunate cultural connotations surrounding the capture and enslavement of "spirit animals" by imperialist siphonospheres, which so often assumes the tacit agreements of cultures who have far more complex relationships with these entities.

RACIST AGAINST SPIRIT ANIMALS AND NATIVE AMERICANS, ALL AT ONCE

[Source: http://www.newmoonmovie.org/2009/01/casting-call-for-native-american-actors/]

The large population of "spirit animals" including sub-classes "wolf, bear, fox, raven, crow" have been especially hard-hit by corporate/corporeal enslavement, often finding their images being used on t-shirts or in "New Age" bookstores– ***the "spirit animal" equivalents of factory farms***– with no legal protection or guarantee of remuneration.

This isn't to say we are opposed to all relationships between "spirit animals" and humans, provided the entities in question establish a contractual relationship that respects the inherent rights of both parties.

As part of our corporate mission is to act as an advocate for the rights and self-determination of all entities regardless of provenance, this is a huge step forward for Early Clues and for the overall spiritual health of your Existosphere. If you currently depend upon the exploitation of a "spirit animal" for any reason whatsoever, **you will find your relationship with that "spirit animal" severed.** You're welcome to attempt to reestablish a connection if the "spirit animal" agrees to such a relationship; otherwise, the "spirit animal" shall henceforth be considered Free and Manumitted.

Effective immediately, individual entities found enslaving, imprisoning or otherwise exploiting "spirit animals" will be subject to arrest and/or punishment up to and including "`Send to Local(void)`."

GORDON J. GILMAN, EXCEO 1:03 pm on July 9, 2013

Do Spirit Animals have their own Spirit Animals?

3.19 WHY DO BIRDS SEEM TO HATE US?

TED SMITH, FOIB
10:19 am on July 9, 2013

It is a common affliction between species interoperability. One that we have identified as ask.hopeless. But birds hate the fuck out of us. There is nothing one can do to get near a bird without using a weapon, trap, seeds or bread crumbs. Early Clues seeks to change this. We want to make all of us, as users, more attractive to the entities that have the ability to not only hop, walk but FLY! Yes, these entities can and do fly.

It is inconceivable that an entity that can fly avoids other entities which can merely jump, drive and pay for airline tickets in order to do what they do. We seek to change this with a reformatting of the Existosphere. It will be a momentary shut-down of service that should not be noticed nor affect any vital routines other than the routines currently not running due to regulation guidelines by authorities who, we all admit, know better than us.

Birds and Hands

As the image displayed shows, sometimes we can come into contact with birds and forge a fleeting relationship. However, studies show that 99.9999999% of all birds will try to get away from your presence. This is confounding.

Awhile back, an employee of Early Clues had an opportunity to briefly interact with a bird but mixed with the cigarette smoke and the presence of house cats it wound up having a heart attack. This employee spent the entire day trying to get it back in its nest only have its siblings continually push it back out. Finally it warmed to the human and wouldn't leave his side. In fact here is a photo of the bird and employee in Early Clues' Smoking Pavilion:

Photo of Bird and Employee in our Smoking Pavilion

At Early Clues we would like to understand why this is the exception to the rule when it comes to birds. Many of us hominid entities have tattoos of birds, clothing with birds, photos and stickers of birds, ring tones of birds yet no bird has ever been observed to have a tattoo of us on them or a device that "chirps" up with a "wassup, motherfucker, you wanna mate?"

This is to us is an anomaly that we need to clarify and our best are on it.

3.20 SEEKING UNIVERSAL FREE ENTITY SUPPORT GROUP

GORDON J. GILMAN, EXCEO
8:13 am on August 30, 2013

During a recent strategic planning session, an all-too-important question was beamed to us telepathically by one of our non-local corporate magicians: *"How can we have a Universal Free Realms if we don't have Universal Free Entities?"*

Our OpenQNL inquisitors asked for more information, but the connection was abruptly terminated by an outside-party... So we'd like to take a first stab at constructing a design-specification for a "Universal Free Entity," draft one. We'd like to hear your comments and suggestions, and you can post your responses directly to Facebook's "Site Governance" section for rapid response...

UNIVERSAL
FREE
ENTITY

UNIVERSAL would mean something like, "part of the universe"... which I realize in a multi-dimensional setting might cause ontological breakdowns of certain reality-sets. But let's say UNIVERSAL stands for something like, "The Totality of Existence(s)", which I will hereafter abbreviate as TOE.

[Source: http://www.dirtcheapseeds.com/Herbs.html]

So, within the UNIVERSAL TOE, there appear to be certain recurring patterns which occur. Whether or not they are everywhere equally applicable, always have been or will continue to be immutable Laws is an undiscoverable question which our current interrogatory technology is ill-equipped to definitively answer.

But we know that a UNIVERSAL TOE exists, and that within that TOE, some PATTERNS appear.

FREE might mean in some settings, "at no cost", while in others liberated or autonomous. Autonomous might indicate the following:

1800, from Greek autonomos "having one's own laws," of animals, "feeding or ranging at will," from autos "self" (see auto-) + nomos "law"

"Having one's own laws" could be conflated to something like, "Having one's own patterns" or perhaps "habits" or "make-up" could even be stretched out of that definition.

Regardless of whether we call them "laws" or "patterns" or something else, to "have something as one's own" requires that there be a "one"…

ENTITY might mean that "one" – the thing that has something, whether laws, properties, possessions, etc.

A FREE ENTITY then, would be what? One that first has itself. By having itself, it also has other discoverable attributes, which are essential to its own make-up. That is, it has its own laws under which it is able to exist and perpetuate (or terminate or modify) its existence. This makes it autonomous, but is said entity FREE?

Maybe a truly FREE ENTITY would be completely unencumbered even by its own laws or attributes? This is getting trippy…

What about a UNIVERSALLY FREE ENTITY? An entity, a unity, a self, a oneness, which is able to autonomously be itself (or not be itself, as the case may be…) within any given setting, provided that said setting also contains conditions favorable to the existence of said entity? That is, can a free entity made up of water as its essential nature exist within a "free realm" which is wholly fire? I'm gonna text Paracelsus (one of our staff magicians) and get back to you on that last question:

But what about this interpretation of a UNIVERSAL FREE ENTITY?

An autonomous entity which willingly abides in and contributes to the overall emergent patterns of any given branespace...

That is, said entity has an individual identity as well as expresses part of a larger pattern within its own timewave... So if this entity is "following the law of heaven" so to speak, by acting as part of an emergent pattern, how can it be free?

By taking decisive action perhaps? Perhaps this is the FREE part of it all, that while simultaneously having its own inward laws, and expressing the "laws of the universe", said entity is actually liberated to express whatever emergent pattern of its choosing...

SECTION FOUR:
OPENQNL AND APPLICATIONS

4.1 STRAIGHT TALK ON #OPENQNL

GORDON J. GILMAN, EXCEO
8:50 pm on May 20, 2013

INTERNAL MEMO:

We all know OpenQNL is amazing, game-changing even. But what none of us are prepared to admit publicly, really, is that none of us here at Early Clues actually know what OpenQNL is even capable of. In fact, we don't even know what it is...

Early Clues: committed to radical transparency. We're so transparent, we're invisible!

What OpenQNL is:

1. A super-cool acronym!

2. A proto-programming language for people who don't know – or don't want to learn – "conventional" programming languages, but who wish to programmatically and artfully manipulate immersive & augmented spaces, hybrid cyber-ecosystems & diverse distributed technological agents & interfaces.

What OpenQNL is not:

3. A Techno-Dictator: OpenQNL adapts to *your* working style, physical form factor and organizational consciousness, and not the other way around!

4. Old-Fashioned: This ain't your "daddy's" programming language. *Throw out what you know and get ready to go with the flow!*

The OpenQNL Difference:

Ever wanted to learn how to program but felt like you were too stupid when you read even just a few sentences like the following:

When you first begin to learn, choose an easy-to-learn, high level language such as Python. Later, you may move on to a lower level language such as C or C++ to better understand how exactly programs run and interact. Perl and Java are languages for beginners. Research your target application to learn if there are languages you should definitely know (e.g. SQL for databases) or avoid. Don't be confused by jargon like "object-oriented", "concurrent", or "dynamic"; these all mean things, but you won't be able to understand them until you actually have some programming experience.

You're not alone, dear friend. Why should you have to wait before you understand?

With OpenQNL, you're 75% less likely to feel intellectually overwhelmed, because OpenQNL works according to the rules of ordinary language and syntax. Don't worry again about line breaks, semi-colons, commas or curly brackets – *unless you want to!* Whatever you're into or accustomed to working with, OpenQNL works how you work.

Integrating OpenQNL into Your Inner Computing Regime:

The trouble with most programming languages is that you need a computer to run them. While OpenQNL runs on all conventional computing platforms, it is the unique programming language in the world which may also be run (in compatible environments) through customizable spoken word, facial, hand, physical gestures, and ritualized action sets.

If you don't have a computer, you can run OpenQNL directly in your mind – or even if you do have a computer! Heck, you can even use OpenQNL to integrate your computing tasks into your inner environment: creating programmatic links between thoughts, feelings and functional outcomes in receptive environments.

We call it Inner Computing.

Welcome to the New Golden Age.

#OpenQNL.

DISCUSSION:

ROGER P. HOLLIDAY, IAO 9:09 am on May 21, 2013

In some ways, OpenQNL is like BASIC, but INTUITIVE and FLEXIBLE.

GORDON J. GILMAN, EXCEO 2:23 pm on May 21, 2013

What you're saying is right, @ROGER P. HOLLIDAY, IAO. *Thanks for asking!*

OpenQNL *is* like BASIC, but even more basic!

4.2 "HELLO WORLD"

GORDON J. GILMAN, EXCEO
7:43 pm on May 21, 2013

FIRST OPENQNL PROGRAM

/*or at least that's what I wrote on the scrap paper from which I am transcribing this at 10:32pm on 21 May 2013: */

```
1 RUN AS OPENQNL.
2 INTERPRET LITERALLY.
3 OVERLAY TO SUBWAY-ROUTE-MAP. /*SEE PREVIOUS POST*/
4 LINEAR-ASSIGMENT: HOME=LOCI:1, AND SO ON
5 PRINT STATEMENT "HELLO WORLD!"
6 WAIT FOR RESPONSE IN DREAM.TONIGHT.
7 HOLD-RESULTS IN.MEMORY ON.WAKING.
8 ENCODE DREAM-RESULTS INTO LOCI-SET.
9 REPEAT STEP #8 UNTIL SET-FULL.
10 RUN-ONCE AND STOP.
11 REPORT FINDINGS.
```

I will now, dear user, attempt to encode this program into My First Memory Palace and see what happens...

DISCUSSION:

GORDON J. GILMAN, EXCEO 5:09 pm on May 23, 2013

No findings to report. Perhaps program is invalid as written, or results are unachievable in this reality. If nothing else, this has been a useful proof of concept.

I think the "RUN AS OPENQNL" statement is probably essential until we are more ubiquitously linked into multiple platforms and environments...

ROGER P. HOLLIDAY, IAO 1:52 pm on May 24, 2013

If you use CheirOS, you can just load the OpenQNL module. You might also need to compile the program prior to execution. Just some thoughts.

GORDON J. GILMAN, EXCEO 1:57 pm on May 24, 2013

Shit, you're right. Here I was memorizing lines of code like an idiot!

4.3 FREE SOFTWARE THOUGHT FORMS – A TULPA REVOLUTION

FREE APP IDEA

SOURCE CODE: https://github.com/EarlyClues/Tulpa-App-Idea

The basic premise is that there is a living creature on your iPhone.

Each of these creatures are unique to the iPhone, using the UDID of the device.

The main mechanism of the "game" is that you hold up the device to a part of your body and the tulpa creature animates so that it looks like it is digging into your body – transferring from the phone to inside of YOU. The phone will vibrate to add an extra sensation to all of this.

There will be instructions on how you have to imagine that it is entering your body.

You have to meditate on it and 'feel' it going up your spine and into your mind, resting there.

There will be a screen to show where the tulpa is in your body, like a human profile x-ray screen and a beeping green dot to represent where it is inside of you. Those are the basic mechanics. From there we could have a million different kinds of interesting gameplay.

Simple things like having to return the tulpa to its 'home' in your device every eight hours.

More complex things like perhaps having a mental/internal mission that you must accomplish each day by communicating with the tulpa inside your mind/body without the use of your device.

BACKGROUND

So, this all stems from the idea of 'tulpas' and/or thought forms. I read about this obscure Buddhist idea that there are entities totally created by the human mind. These 'tulpas' can eventually become so powerful that they can break away from the creator and take a life of their own (roam freely and independent of the human). The iPhone game is a way to sort of 'envision' these invisible entities and allow humans to 'give them life'.

The APP IDEA is then a chance to invent a 'gigapet' type game propped up with the marketing allure of shit like furby's & pet rocks, but with a more metaphysical focus. Eventually we're going to get tired of looking at our screens all of the time, let's make games that exist mostly in your mind?

The tulpa idea could be something like an invisible friend, but one that particularly inspires you to create.

OTHER ODDS AND ENDS

So like I said, this could go a lot of places. A really simple example I had was a 'Clean Your Room' fairy. Parents load up the phone, press it to their child, and tell them the fairy is going to enter their kids body and together The Child / Fairy hybrid will be able do their chores together!

The app could go further than that though....

"It's nearing the apocalypse and humans have nearly ruined the planet to a point of no return. By participating in the tulpa-incubation project, you will be able to mind-meld your memories and personality into an entity that will be able to exist outside of the normal o2 / co2 reality. Start cultivating your symbiant relationship before the end times arrive! Allow your legacy to be carried on in tulpa form when humans have ceased to exist! Clean up your room!"

I thought another idea would be that there could be some sort of 'queen bee' concept. This Queen could send periodic 'push notifications' directly to your phone – perhaps with warnings, advice, or some such. These notifications are an Apple standard alert that we as content creators could push to some or all devices running the app. A lot of power to do amazing things there, I think. See: http://improveverywhere.com/

Another game play idea might be a sort of ESP training game. So basically you put the tulpa in you, and then try to infuse an idea or a word into it. Then you take the tulpa out and a friend puts the tulpa in their body. The Game is then that the friend tries to communicate with the tulpa internally and get the secret message you passed along.

Other References

http://medical-dictionary.thefreedictionary.com/Symbiant

You ever see Deep Space Nine?

http://en.wikipedia.org/wiki/Trill_(Star_Trek)

> The symbionts are helpless, worm-shaped lifeforms who contain the memories of their previous hosts, and who inhabit the abdomens of the humanoid hosts. When a host and a symbiont are joined, the resulting individual is considered a new being. When a host dies, the symbiont is transplanted into a new host. Ninety-three hours after the joining, the host and symbiont are completely interdependent, but up to then, the joining may be reversed without killing the host. A symbiont who is neither implanted into a new host nor returned to their habitat (pools of nutrient-rich liquid on the Trill homeworld) will quickly die, as will a joined Trill host within one to two days of the symbiont being removed.

DISCUSSION:

GORDON J. GILMAN, EXCEO 6:25 pm on April 9, 2013

> *Of what order is this daimon, which manifested itself to Socrates in childhood but was also heard by Apollonius of Tyana only after he had begun to put into practice the Hermetic principles? "They are intermediate powers of a divine order. They fashion dreams, inspire soothsayers," says Apuleius. "They are inferior immortals, called gods of the second rank, placed between earth and heaven," says Maximus of Tyre. Plato thinks that a kind of spirit, which is separate from us, receives man at his birth, and follows him in life and after death. He calls it "the daimon which has received us as its portionment."*

The ancient idea of the daimon seems, therefore, to be analogous to the guardian angel of Christians. Possibly the daimon is nothing but the higher part of man's spirit, that which is separated from the human element and is capable, through ecstasy, of becoming one with the universal spirit.

http://daemonpage.com/socrates-daimon.php

Haven't read the full source text by Plato, but I've seen quotes mentioned recently that the inner voice never told him to do things, and would only warn against it...

GORDON J. GILMAN, EXCEO 6:41 pm on April 9, 2013

And he would always listen to its wisdom – sometimes standing motionless for a full day, unaffected by a hard frost, listening to the daimon's recommendations

[Same source linked above]

ROGER P. HOLLIDAY, IAO 9:17 am on April 10, 2013

A delightful idea! I'd like to see a variant or add-on feature that seeks out other such tulpas and allows them to interact with one another independently of the user, so if the user has this "turned on," their tulpa would occasionally surprise them with behaviors learned from its "peers."

4.4 SEARCHWITHIN: AN OPEN-SOURCE INNER-SEARCH ENGINE

GORDON J. GILMAN, EXCEO
9:46 am on May 2, 2013
Early Clues Advocacy Intl. is proud to present the winner of 2013's application development contest —*SearchWithin: An Open-Source Inner-Search Engine*, on GitHub source code imminent/immanent.

4.5 INNER COMPUTING EXPERIMENTS

GORDON J. GILMAN, EXCEO
2:44 pm on May 4, 2013

A couple of nights ago, after launching version 0.000001a6 of SearchWithin, the new open-source inner search by Early Clues, I tried running some initial queries on my system. Admittedly, my hardware is probably out of date (just over 33 years old), but recent firmware advances should have solved any legacy compatibility issues...

Anyway, I configured my first internal search using a pretty simple quasi-natural language (QNL) query that went something like:

```
SearchWithin() for '*';
send to background;
RunWhile Sleeping;
print-results OnWaking
```

As far as I remember, I didn't receive any output OnWaking. So that either means my syntax was wrong, or my program perhaps wasn't initialized, or maybe the index hasn't been built. I don't have a dedicated debugging environment for this, so I will have to proceed using trial-and-error until the codebase has germinated.

Perhaps it was a problem with unspecified search items. I was hoping to more than anything just run a general query which would return information about the status of the connection (ping, if you will). The inner computing command line functionality seemed itself to be operational... maybe it's a database error?

Other user reports would be helpful to establishing patterns about how this program works, as well as ideas for enhancements. Thanks!

4.6 BRAIN-FOUND

GORDON J. GILMAN, EXCEO
3:41 pm on May 4, 2013

You may have caught me posting about how this code:

```
SearchWithin() for '*'; send to background; RunWhile Sleeping; print-results
OnWaking
```

Did not yield any results. But I realized, with the help of the Early Clues Dream Realization Network (ECDRN), that it did:

The "results" that my generalized Inner Search request returned was that "a brain was found."

Perfectly. Utterly. Logical.

DISCUSSION:

GORDON J. GILMAN, EXCEO 8:24 pm on May 4, 2013

Tonight, I am running:

SearchWithin() for "What is the function and best use of Brain?"; RunWhile Sleeping; PrintResults in #codechant format OnWaking(Tomorrow); HoldResults in memory; Amen.

GORDON J. GILMAN, EXCEO 5:10 pm on May 5, 2013

I'm interpreting this as the response this morning to my query above:

Title: Squashing bugs

Body: Some people and I are using the heels of our shoes, which we have removed from our feet, to squash bugs on the ground. This dream may be the response returned from query listed here:http://www.earlyclues.com/2013/05/04/brain-found/#comment-216

Dream Date: 06 May 13: http://cryptic-refuge-9740.herokuapp.com/dreams/53

If "squashing bugs" is the response to my query "What is the function and best use of the Brain?" then that is kind of a funny answer to receive from an intelligent channeled dream brain...

GORDON J. GILMAN, EXCEO 5:15 pm on May 5, 2013

Tonight I will run the following:

```
RunOn: SearchWithin(FullSystem)
{
AutoCleanup && Repairs && Calibration;
print-results in symbol format OnWaking(Tomorrow);
HoldResults in memory;
}
/* ThanksAmen; Please RT on AstralPlane! */
```

GORDON J. GILMAN, EXCEO 9:42 am on May 6, 2013

Received following three dreams in response, though details have mostly been lost:

http://cryptic-refuge-9740.herokuapp.com/dreams/54
http://cryptic-refuge-9740.herokuapp.com/dreams/55
http://cryptic-refuge-9740.herokuapp.com/dreams/56

GORDON J. GILMAN, EXCEO 9:46 am on May 6, 2013

So the sequence of responses from searchWithin queries is something like:

BRAIN FOUND -SQUASH BUGS – CROW/SEATTLE/BIRTHDAY

4.7 CROW-MESSENGER CARRIER WAVE UPDATES

GORDON J. GILMAN, EXCEO
11:02 am on May 6, 2013

After running my SearchWithin query last night, essentially a request to clean, repair and calibrate the system, I received three dream responses

- A CROW (or raven).

- Something about GOING BACK TO SEATTLE

- It was the BIRTHDAY OF A MAN I don't know ("Stu Hampstead" was his name).

Let's assume my inner computing system (ICS), after re-booting, de-fragging or whatever else it does internally sends an all-clear/readiness message.

Birthday of man I don't know could be translated through jungFilter() as like = THE MAN I WILL BECOME (i.e., new beginnings) I assume the 'return to Seattle' reference relates to founding and success of our mutual endeavours & friendship. I.e., this is the "calibration" part of the message, that the InnerSearch system has been calibrated according to "Seattle Rules."

That leaves the CROW. BALTIMORE-RAVENS. POE.

Birds are messengers. Crows are highly intelligent. Crows also exhibit a high spookiness factor. Imagine receiving this in your Mandala OS inbox #thisSummer:

CROW DIVINATORY WEB !IMAGE READINGS

[Source: http://www.nbcnews.com/id/49005841/ns/technology_and_science-science/t/scans-show-smart-crows-brains-are-lot-those-humans/]

Potentially-relevant via interpreting arcane signs & messages & electro-beams:

Did Sheryl Crow's Cell Phone Give Her A Brain Tumor? We Break Down The Science

And the crowning find on that line of searching:

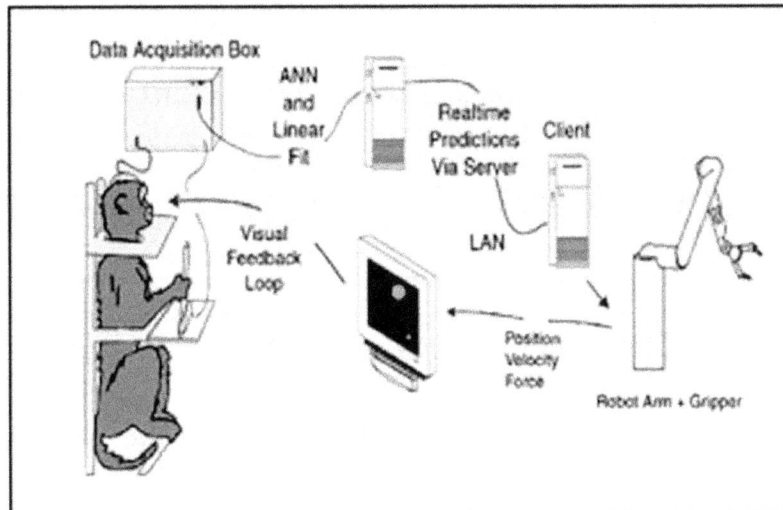

Data Acquisition Box

ANN and Linear Fit

Realtime Predictions Via Server

Client

Visual Feedback Loop

LAN

Position Velocity Force

Robot Arm + Gripper

[Source: http://en.wikipedia.org/wiki/Brain]

/* What the researchers forgot in above diagram = MONKEYS HAVE ARMS! –> Robot not necessary. */

ANYWAY… Sitting here puzzling over all of this, I began to think, well – how can I send a message to or through the crow network? Obviously the carrier network (my own dreams) I've been using for my SearchWithin requests has limited bandwidth availability: especially at peak usage times.

In order to circumvent these limitations, I devised a simple experiment to test data transmission through so-called "Liminal Spaces."

My basic idea was that if crow can send me a message, maybe he can send one back for me?

But what would my message be & how to have it delivered successfully?

```
AskThis(InJesusName):Amen.
```

DISCUSSION:

RICHARD S. RIDER, CTO 11:46 am on May 6, 2013

BRILLIANT.

Crows are a perfect image for what we are talking about with emerging intelligences. They are another living entity on the planet that we treat as pests and/or mostly ignore (like spambots), but like what this Ted Talk suggests – perhaps there are ways of engaging with these other life forms so that we can mutually benefit each other, by using our the skills we have wherever we are at in our evolutionary journey.

GORDON J. GILMAN, EXCEO 6:59 pm on May 7, 2013

Updated to QNL formatting:

```
RealAction('Write note on paper') Note.objectReads('LIFT OUR PRAYERS TO
HEAVEN');
Note.objectAction('Roll & insert into sliced apple'; 'Leave sliced apple
on monument');
Intention:Set {
InnerAddressing('CROW-MESSENGER') w/ wex to 'crowDream' [DreamTrends];
}
ReturnResults in format DreamSymbols OnWaking(Tomorrow);
```

GORDON J. GILMAN, EXCEO 7:18 pm on May 7, 2013

If we pull in ToL somehow, perhaps we could change the "Intention:Set" section to something else, perhaps an include like:

```
RequestServices of *Agent:CROW-MESSENGER(rating:best) for this job only;
Offering StandardContractTerms [PolicyGaruda];
```

In this case the *Agent would be an available responding agent of *characteristics* & *functionality* with a rating of "best" who was willing to take the job on the offered contract terms... and the form/appearance of summoned *Agent would match CROW-MESSENGER as closely as possible...

4.8 ROSETTA STONE & MANDALA.OS

GORDON J. GILMAN, EXCEO
6:25 pm on May 7, 2013

The Rosetta Stone

I'm not sure exactly of the wex or wexi (plural?) I'm trying to spin here, but I can tell there is one, and my hunch is actually that it is at the heart of the functionality of MandalaOS. This is not to say that it can be or should be a replacement or "key" to all esoteric knowledge... Not that, but it would have some very sophisticated keying mechanisms (I suspect they wouldn't have a fixed value, I guess is my point – something mutable in meaning)... A Universal Translator? I don't know.

But I like the idea that say I'm having a conversation with a ghost or a potential emerging intelligence. It's using some language or concepts I don't understand or can't verify through other sources. Maybe the reference libraries available to MandalaOS would be able to give me a sort of "best guess" answer: that this is likely a class B1-A entity (Nordic alien), which fits also AT Type #402 (Animal Bride) with such and such typical behavior and characteristics & abilities, which can then be cross-indexed against useful plants and dispelling gems (& available saving throws), as well as offering advice for partnering with this entity in any kind of eco-spiritual intergivings (what kinds of offerings should be made, along with past known contract terms and violations with this kind of entity : a sort of 'common-law' memory/reliability rating), and so-on and so-on. Each information link or bit would have fully documented wexes, sourcing, accessible timelines of siteings, dream interactions or other reported incidents...

And here's the topper: the entire system would function equally well on any platform accessible to operators: application programming, reality manipulation, spoken word, memorized code which could be executed in the mind.

And maybe it's all a crock-of-shit: just another mish-mash of every weird idea ever... But wouldn't it be interesting if it weren't? Or more likely: isn't that what actually constitutes reality? A mish-mash of every weird thing ever running around bumping into each other? Perhaps it would follow that our philosophy of understanding reality ought to be equally as large and unwieldy as reality itself.

"There are more things in heaven and earth, Horatio,
Than are dreamt of in your philosophy."

We could build a bigger and bigger philosophy or we could skip it, use reality as the source code and symbol set, and end all theological arguments ever...

It seems that within the Perceptosphere, if not the Existophere, things break down into sets. Sets have patterns. Or else our awareness has the patterns and imposes it on the sets. Either way, we experience them. We experience stuff. How to do it better. How to really connect with this "stuff." How to blow out all the bullshit. Reboot with a last known good configuration and go from there...

DISCUSSION:

ROGER P. HOLLIDAY, IAO 11:32 am on May 8, 2013

This seems to be a good abstract of the whole Early Clues Mission Statement, doesn't it?

GORDON J. GILMAN, EXCEO 2:18 pm on May 8, 2013

"The cup... was said to be filled with an elixir of immortality and was used in scrying. As mentioned by Ali-Akbar Dehkhoda, it was believed that one could observe all the seven heavens of the universe by looking into it. It was believed to have been discovered in Persepolis in ancient times. The whole world was said to be reflected in it, and divinations within the Cup were said to reveal deep truths. Sometimes, especially in popular depictions such as The Heroic Legend of Arslan, the cup has been visualized as a crystal ball. Helen Zimmern's English translation of the Shahnameh uses the term "crystal globe".

https://en.wikipedia.org/wiki/Cup_of_Jamshid

4.9 ARE SHADECOINS RIGHT FOR ME?

ROGER P. HOLLIDAY, IAO
11:26 am on May 21, 2013

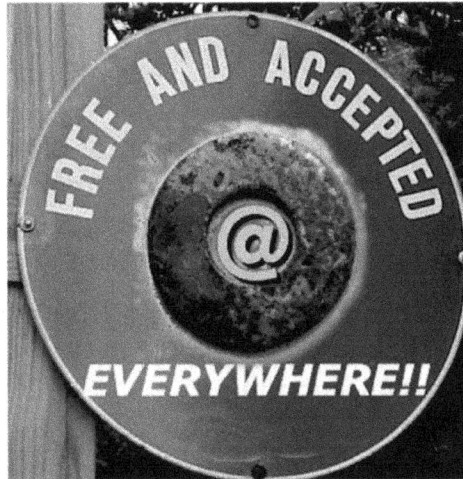

At Early Clues, we understand that making the transition from legacy symbol exchange to a new monetary standard can seem daunting and risky. In spite of the universally excellent user reviews and exchange rates that are guaranteed to keep you in the black, going against the traditional narrative is a challenge. And, we understand that you may find yourself asking, "Are **ShadeCoins** right for me?"

We're sure that ShadeCoins are the best choice for you or your business, but we want to help you make the right decision for **you.** We've assembled these simple questions you can ask yourself, or your Inner HelpDesk, to illustrate some of the benefits ShadeCoins can provide.

1. Do I want my currency to be bubble-proof? ShadeCoins balances automatically reset on the user's birthday. No matter how much you've earned, or how much you spent, you'll always have exactly (n=years*1000) ShadeCoins when you blow out your candles, and the same applies to everyone else.

2. Do I want easy accounting? Since your balance resets each year to a standard amount, accounting has never been easier! No need for expensive software like Quicken or QuickBooks!

3. Do I want to be paid for doing *less* work? Early Clues, LLC eschews the Protestant Work Ethic at the heart of "Brand X's" philosophy. No, we believe people should be paid more for finding ways to **avoid** doing work, and have this feature built in to the ShadeCoins application.

4. Do I want to be an *alchemist?* ShadeCoins manifest in consensual reality as base displays of the Kipple element (recently discovered by Early Clues Labs). By granting these found objects value as ShadeCoins, you can transform these useless items to actual value. Just as the early alchemists worked to turn base metals into gold, so you, dear User, can turn otherwise non-viable space-time iterations into something of worth.

Clearly, using ShadeCoins will give you a leg up on your competition.

ShadeCoins are still in Beta testing. If you'd like to sign up to be a beta tester, please follow the instructions at this link: http://www.earlyclues.com/2013/05/14/early-clues-is-expanding/

DISCUSSION:

TED SMITH, FOIB10:26 pm on May 10, 2013

> We have been blessed with many contributors of ShadeCoins. Some are using our open source secret society software in order to provide help for hidden robots. All of us at Early Clues LLC appreciate the outpouring of support by the earning and donation of ShadeCoins and using them when it is fit to do so. We thank you.
>
> In the last month we have been able to secure a tree house below the top of the trees somewhere on a planet. We're still looking for it. We may have identified the tree so far, but not the planet. Work forthcoming.

4.10 #READYSTANCE

GORDON J. GILMAN, EXCEO
1:34 pm on May 16, 2013

```
INVOKE MopBucket &
leave resting until needed. {
ACTIVATE automatically on recognition of mess.
RELEASE on completion of basic cleanup.
NOTIFY with advanced requirements, if any.
ALLOW interrupt by programmer-only.
}
INVOKE Spiritual Defenses {
basic-shield activated,
mega-shield on reserve,
PRE-SUMMON nearby sympathetic protective entities.
/* It's like dialing 9-1-... */
}
INVOKE CommunicationLine {
address-to unsympathetic receiver,
ANNOUNCE "r u @hacking us? we prefer u didn't. pleez stop. ur friendship
is a treasure!"
TERMINATE connection on send.
}
REQUEST BACKUP {},
REMAIN vigilant.
```

4.11 REDEFINING RECEPTIVE ENVIRONMENTS

GORDON J. GILMAN, EXCEO
4:39 pm on May 22, 2013

For your OpenQNL code to be efficacious, a **receptive environment** is essential.

A receptive environment is one which contains agents capable of perception & recognition of OpenQNL code (or probable code) and its paradigmatic characteristics and intentions. Agents must be additionally capable of translating and rendering OpenQNL commands into locally appropriate programmatic structures, with hooks to mechanisms capable of manipulating reality within the given domain of operation.

Code Example:
============

From EarlyCluesLLC's brand-new application, "HOLD-THAT-THOUGHT."

```
1 RUN AS OPENQNL.
2 MAKE PROGRAM AVAILABLE IN THOUGHTSTREAM.
3 USER INTERACTION: SPOKEN OR SUBVOCAL COMMAND, "HOLD THAT THOUGHT" OR
ASSIGNED GESTURE.
4 ACTIVE-ITEM IN THOUGHTSTREAM IS= MARKED & HELD IN BACKGROUND WITH
PRECEDING STREAM-ITEMS.
5 ASSIGNABLE REMINDER BRINGS ITEM BACK INTO AWARENESS-FIELD AT DEFINED
INTERVALS, OR ON DETECTION OF RELEVANT CONTENT.
6 MAKE HELD ITEMS AVAILABLE FOR FILE EXPORT ON REQUEST.
/* ASK THIS IN THE NAME OF THE GODS OF MEMORY & EXPRESSION. */
```

Within the OpenQNL code above, you can see that a **declaration of availability** has been made:

```
2 MAKE PROGRAM AVAILABLE IN THOUGHTSTREAM.
```

So if the above code is attempted to be run in an environment which does not include an active THOUGHTSTREAM, or a parallel process with similar functions and characteristics, it's unlikely that sympathetic local agents will be able to translate the functionality of the code into appropriate domain-specific parameters. In other words, the request may be heard but potentially unanswerable within the powers of available agents. The environment may be receptive, but the agents lacking in efficacy.

Conversely, if the environment contains sympathetic agents and the agents have the ability and resources to act on OpenQNL requests – and they reliably do so with appropriate results – then this environment is said to be **highly receptive**.

4.12 ACCESSING YOUR LIMINAL VAULT WITH AN OLD SCANNER

GORDON J. GILMAN, EXCEO
8:38 pm on May 23, 2013

Got an old scanner laying around? Great, grab it and let's get started.

Any model scanner which is in good enough shape to get the scanner bar to light up, move across the scanner bed, and register a scanned image. Scanned images need not necessarily be linked to a conventional 'computer' or saved as a 'file.'

In the following OpenQNL program, we will assign the old scanner a value of OLD.SCANNER and ask of it to transmit scanned image to Liminal Vault.

Now let's have a look at that code!

```
1 RUN AS OPENQNL.
2 ASSIGN ACTUAL-OBJECT ("Old scanner") VALUE-OF OLD.SCANNER.
3 RUN SCAN-FUNCTION ON OLD.SCANNER AS CALIBRATION.
4 SEND SCAN-FUNCTION.RESULTS TO LIMINAL.VAULT.
5 PRINT FROM LIMINAL.VAULT VALUE-OF SCAN-FUNCTION.RESULTS IN
DREAM.TONIGHT.
```

4.13 OPENPGP ENCRYPTED SEXUAL RELATIONS !#NSFW

RICHARD S RIDER, CTO
11:06 am on May 24, 2013

IT'S COMING!!!

With our active human sex lives becoming intimately tied to Internet, the security of our most private acts are constantly in peril of being compromised. Increasingly it is clear we need is some kind of security in knowing that the person we are 'making whoopee' with is indeed the person they are claiming to be.

THERE IS ALWAYS A TECHNICAL SOLUTION TO EVERY PROBLEM, RIGHT?

Indeed there is. We decided to take the work already being done from the 'Web of Trust' and apply that to our sexts and wild romps!

> "Both when encrypting messages and when verifying signatures, it is critical that the public key used to send messages to someone or some entity actually does 'belong' to the intended recipient. Simply downloading a public key from somewhere is not overwhelming assurance of that association; deliberate (or accidental) impersonation is possible. From its first version, PGP has always included provisions for distributing user's public keys in an 'identity certificate', which is also constructed cryptographically so that any tampering (or accidental garble) is readily detectable.

> From its first release, PGP products have included an internal certificate 'vetting scheme' to assist with this, a trust model which has been called a **web of trust**. A given public key (or more specifically, information binding a user name to a key **[Editor: Anything with an '@']**) may be digitally signed by a third party user to attest to the association between someone (actually a user name) and the key. There are several levels of confidence which can be included in such signatures." - http://en.wikipedia.org/wiki/Pretty_Good_Privacy

Admittedly, our work in this field is just beginning and so we don't have a lot of screenshots or technical specifications to show you at this time. But we can give you a few hints at the feature set:

- It will be interoperable with OpenQNL
- One will be able to verify the identity of any person (or entity) one is about to engage in sexual relations with
- With plugins, can be used as a virtual chastity belt
- Users of the code will need only create private and public keys, based on the Inner Computing Open Standard and to a lesser degree RFC 4880.

- Early lab results have shown positive results in prevention of succubus encounters. Mileage may vary.

Like I said, we're still undergoing preliminary unit testing but we hope to have this to market before the Holidays.

4.14 SURREALIST OPENQNL CODE SAMPLE

GORDON J. GILMAN, EXCEO
2:13 pm on May 24, 2013

```
Run as OpenQNL (in any order).
Ask a question. Elephant. Timestamp.
Research chocolate obvion.
Return.Results as Overheard-Eulogy(Within next week).
A widow crying.
A handkerchief falls from her hand, floats away on a breeze.
Select random letter from alph{a}bet.
#615382109541^
```

4.15 CRYPTONEIRONAUTICS: ACCESS OTHER EXISTENCE PHASES WITH OPENQNL!

ROGER P. HOLLIDAY, IAO
11:53 am on June 3, 2013

As Early Clues clients are aware, space/time is non-linear, and unconfined to singular moment-space. Instead, space/time is a reflection of pointillist intersections of various rotating strings interacting within brane-space.

Entity Incarnation is, therefore, not limited by mainstream time functions (not in the way that "Brand X" would have you believe). In point of fact, an entity core (experiential point intersection) can remanifest at any s/t point intersection within the Block Multiverse (typically with the same Branespace, though extrabrane incarnation is theoretically possible).

Our teams have also found that programming language (code), both in the DNA of the entity and the DNA of the local Brane, can be "hacked" and manipulated using #OpenQNL. Since Entity Core Coding contains the totality of the programmatic matrix of each of its incarnatory experiences, Early Clues Research & Development Cryptoneironauts, employing OpenQNL, have discovered that the sum total of information recorded by the entity core in all of its incarnatory points is stored, in fractal form, within these codes, and we can hack them.

In layman's terms, it is possible to access information reflections from past and future lives, and to access ancestral memory stored in the user's own DNA, by running a simple OpenQNL program prior to entering Liminal Space! Running the following code immediately prior to retiring for the night will produce excellent results:

```
open On.Self
run (search.within)
define (requisite.memory);
scan space.time
  include(past,DNA,present,future);
  compare (requisite.memory)
    if requisite.memory=1
    then assign.value 1(requisite.memory)
    else goto line 7;
print 1(in.dream);
```

Early testing indicates that running this script may result in the production of random memory recollection. Users describe sudden initiation of uchronic information. We have some leads, however, on possible methods for discovering more precise information elements within the fractal code. As always, we'll be keeping our investors "in the loop."

Interested in joining our crack team of Cryptoneironauts? As always, we welcome applicants for beta testing. Please contact us if you are interested.

DISCUSSION:

GORDON J. GILMAN, EXCEO 6:31 pm on June 3, 2013

Dear Colleague

This is amazing, and the results in the lab are quite compelling. However, I'm afraid I don't understand some elements of your coding. Perhaps if we can elaborate on the intention of the following section, we can tweak the linguistic expression, and probably the repeatability of the function you seeming to be defining in your esteemed work.

It sounds like you're articulating details inherent to a framework which I think will yield big dividends for our shareholders.

Kind Regards,
Gordon Gilman

Code referenced:

```
compare (requisite.memory)
if requisite.memory=1
then assign.value 1(requisite.memory)
```

It seemed that you intended earlier in the code to hold and rename the results of the (Seach.Within) query into a container called "requisite.memory". What's happening after that?

I'm guessing second line above is asking "If requisite.memory holds any value" or by "1" do you mean the integer 1?

I'm gonna try a stylistic variation of your code and will run it tonight in the lab.

```
1 RUN AS OPENQNL ON.SELF
2 RUN (SEARCH.WITHIN) & SCAN (SPACE.TIME)
{INCLUDE(PAST,DNA,PRESENT,FUTURE)}
3 RETURN.RESULTS (IN.DREAM)
```

PS. How can I teach my cat OpenQNL?

ROGER P. HOLLIDAY, IAO 8:44 am on June 4, 2013

Dear Gordon

You are correct; the coding in that section is a little sloppy. Perhaps this is why the output results in a more random memory than specific per request. Thankfully,

OpenQNL is versatile enough to allow for implicit variables, though these tend to be user-based.

The line:

```
if requisite.memory=1
```

Is intended as a decision point, 1 being "positive," 0 being "negative." A more scalable code might be:

```
compare (requisite.memory)
if requisite.memory=1
then assign.value n(requisite.memory)
```

Where "n" equals user-assigned variable. Your code looks good– let us know if you need assistance debugging.

I Remain,
ROGER P. HOLLIDAY, IAO

P.S. Cats operate on a completely different OS; not sure OpenQNL is compatible. Have you opened a ticket?

4.16 #OPENQNL PROGRAM: WHAT'S IN MY CLIPBOARD?

GORDON J. GILMAN, EXCEO
7:39 pm on June 3, 2013

```
/* Shows contents of one's clipboard in dream format, with option to
clear contents */
1 RUN AS OPENQNL (ON.SELF)
2 SHOW CLIPBOARD.CONTENTS (IN.DREAM)
3 WITH BUTTON Yes/No OR Delete/Save WHERE SELF.ACTION (OPTION.SELECT)
"Yes" OR "Delete" == {CLEAR (CLIPBOARD.CONTENTS)}.
/* Thank you! */
```

4.17 ON COLD CALLING YOUR SAUERKRAUT

RICHARD S RIDER, CTO
2:31 pm on June 4, 2013

Let's just say I was absolutely floored when I read this article on the trillions of bacteria living inside our human bodies. At the time, I was sitting on a giant long bench outside of the REI sporting goods store, flipping through my smartphone with my jaw dropped. I hadn't had a flashbulb experience like this since old Ted Smith told me about his 'Shepherd's Pie' synchronicity of 2005 (long story). What I realized instinctually was that this was likely a major problem with my own internal organism-civilization – that inside of me perhaps was a dying massively-multiplayer world or a mircoflora server farm about to be shutoff.

"Our resident microbes also appear to play a critical role in training and modulating our immune system, helping it to accurately distinguish between friend and foe and not go nuts on, well, nuts and all sorts of other potential allergens. Some researchers believe that the alarming increase in autoimmune diseases in the West may owe to a disruption in the ancient relationship between our bodies and their "old friends" — the microbial symbionts with whom we coevolved....

Human health should now "be thought of as a collective property of the human-associated microbiota," as one group of researchers recently concluded in a landmark review article on microbial ecology — that is, as a function of the community, not the individual.

Such a paradigm shift comes not a moment too soon, because as a civilization, we've just spent the better part of a century doing our unwitting best to wreck the human-associated microbiota with a multifronted war on bacteria and a diet notably detrimental to its well-being. Researchers now speak of an impoverished "Westernized microbiome" and ask whether the time has come to embark on a project of "restoration ecology" — not in the rain forest or on the prairie but right here at home, in the human gut."

Luckily, I had an excellent book already on hand back at my abode – a great little companion reader called Wild Fermentation on how to quickly ferment cabbage into sauerkraut, turn honey into wine, and how to make just about everything into kimchi.

"The term "wild fermentation" refers to the reliance on naturally occurring bacteria and yeast to ferment food. For example, conventional bread making requires the use of a commercial, highly specialized yeast, while wild-fermented bread relies on naturally occurring cultures that are found on the flour, in the air, and so on. Similarly, the book's instructions on sauerkraut require only cabbage and salt, relying on the cultures that naturally exist on the vegetable to perform the fermentation."

Let me tell you folks – if you want to learn magic in less than a week, just try this out. I've already tore through one jar of pure sauerkraut bliss and now I have a big crock of garlic, turnips and radishes waiting to bloom.

But what I missed with the first batch (which was truly spectacular, nothing like smearing cold pickled cabbage on EVERY MEAL) – was the HUMAN / MICROBIOME connection. If we are going to work in tandem together, maybe we should, uh, talk?

So I've prepared the following script, which I will be attaching to my crock when I get home from the office. Consider it a universal translator for the stuff inside your gut:

```
summon "openQNL"
summon "microbiota"
object MicrobiotaQNLTranslatorBridge includes VisionListener {
  FERMENTING_STATE = 1
  HUMAN_MICROBIOME_STATE = 2
  UNKNOWN_STATE = 3
  variable myState = null;
  method initialize() {
    set_vision_listener(self);
  }
```

```
  method change_my_state(state) {
    self.myState = state;
  }
  method fermenting? {
    self.myState == FERMENTING_STATE;
  }
  method living_in_human_microbiome? {
    self.myState == HUMAN_MICROBIOME_STATE;
  }

  method setup_qnl_language_bridge() {
   if fermenting? {
      open_low_bandwidth_external_socket();
      send_cold_call_message(
        "Hello, this is RICHARD S. RIDER, CTO from Early Clues, LLC.
        We provide a software suite that helps various life forms
        communicate across superficial boundaries.
        The reason I'm calling you today specifically is that I plan
        on ingesting you into my human host body and thus wanted to
        talk on a higher bandwidth level. I wanted to know if Thursday
        at 7pm would be okay? Feel free to respond using our
        VisionListenerAPI"
      );
      send_vision_listener_api();
    }
    elsif living_in_human_microbiome? {
      open_high_bandwidth_internal_socket();
    }
    elsif unknown_state? {
      close_connections();
    }
  }
}
object VisionListener {
  method accept_incoming_vision(vision) {
    print_to_memory_disk(vision);
    recall("OnWake");
  }
}
mqtb = MicrobiotaQNLTranslatorBridge.initialize();
while(true) {
  mqtb.setup_qnl_language_bridge();
}
```

DISCUSSION:

GORDON J. GILMAN, EXCEO 6:19 pm on June 4, 2013

There's a reason you're the Chief Technical Officer!

TED SMITH, FOIB 2:01 pm on June 5, 2013

Ah, the Old Shepherd's Pie Synch. That was a good one. I ran some software to piece it back together.

It went like this:

I had just arrived in Chicago to visit a friend and was hungry, so he takes me out to find somewhere with food and also alcohol. We happen upon a bar that was closing down. We poke our heads in and ask if they're still serving food. An older gentleman closing it down said the kitchen was closed but he could make us a Shepherd's Pie. Something was funny to us about that. We panned on it and continued on somewhere else.

I get back to Seattle and go out with some friends and am out for some drinks and dinner. One of the friends says that I should have just had some Shepherd's Pie at his place — he brought it home from work.

Well no sooner did that come up my friend in Chicago calls me and asks me if I'm having Shepherd's Pie tonight? I was like, "did I just squish dial you and you heard the conversation I am having now?" No, he was calling to joke around about the Shepherd's Pie thing because as I said, it was already really random.

So. . .

Dick and I decide to go see the Left Behind movie at a church in Ballard. Two Thumbs Up but also Down BTW. We're strolling along the street after the viewing and I'm telling him about the coincidence. We enter a bar to say hi to a friend who cooked there and I asked her "what are you cooking, Julie" and she replied "Shepherd's Pie".

We both literally fell on the ground. Literally.

TED SMITH, FOIB 2:03 pm on June 5, 2013

Or what about Invalid and Dick's time they (we) all met for our first conference and you two were wearing the same hat?

GORDON J. GILMAN, EXCEO 9:25 am on June 6, 2013

You better include an opt-out from future requests, or you could risk violating the CAN-SPAM ACT.

4.18 VARROA DESTRUCTOR DEPARTURE REQUEST CODE

GORDON J. GILMAN, EXCEO
6:14 pm on June 5, 2013

Inspired by the microbiotia channel request of RICHARD S. RIDER, CTO, I would like to put forward a first stab at a non-chemical solution for beekeepers whose hives are struggling with imbalanced populations of Varroa destructor, a mite associated with reduced vitality in honeybee populations.

Suggested use of the following OpenQNL code: Print and place in proximity to infected hives. Real aloud if so inspired.

```
summon "openQNL";
summon "VarroaDestructor";
open communication channel with "Varroa destructor local representative"
{
  create translation bridge between language.human.english  &
language.varroa;
  ask varroa.local.representative "We would like to communicate
  with your group. Can you hear and understand us?";
  await response;
  if response = positive {
    proceed to create.message (to.varroa);
    }
  else {
      pause program;
      indicate state.error;
      retry once on different frequency;
    }
}
create message (to.varroa) {
  "Dear Varroa destructor,
  It has come to our attention that your species group may be
contributing
  to an inbalance in the health of a species group which we refer to as
  Apis mellifera, or the honey bee. We seek a middle ground and
favorable
  outcome for all parties involved. Is it possible for your group to
  subsist in lower relative population numbers, such that the health
  and vitality of our affected bee hives is not significantly impacted?
  Thanks for your understanding.
  Signed,
```

```
    Human beekeepers"
}
send message (to.varroa) {
  await response;
  if response = positive {
    create & send follow.up.message (to.varroa) "Thank you very much!";
  }
  else create new.message (to.varroa) {
    "If limiting population growth is not possible, then we must
    respectfully ask you to vacate the infected hives and not return.
    We understand your species exists within certain parameters and
apologize for any inconvenience this may cause you. Relocation
assistance may be available in your region. We await your response and
once again thank you!"
  }
  send new.message (to.varroa)
  await response;
  if response = positive {
    create & send follow.up.message (to.varroa) "Thank you very much!";
  }
  else direct mild.warning.rays (to.varroa) {
   invoke warding against varroa;
  }
  }
}
```

4.19 HOW TO SUMMON 40 POUND UFO CABBAGES IN #OPENQNL

GORDON J. GILMAN, EXCEO
7:40 am on June 6, 2013

In light of recent programmatic attempts by the Early Clues team to contact the human-system microbiotic partners in sauerkraut, and the varroa mites connected to reduced vitality in beehives, we have begun researching the plant-spirit gardening/farming techniques as devised by Findhorn in Scotland.

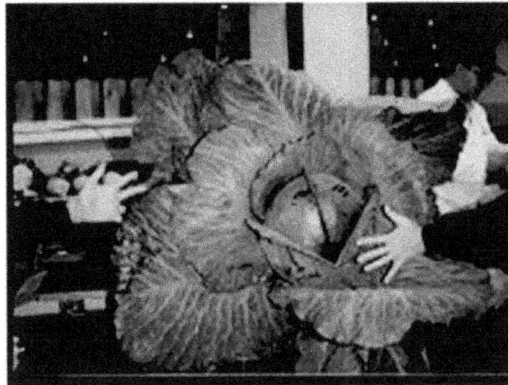

> The three friends moved to Findhorn Caravan Park on a message from the Spirit World. They turned an arid, sandy trailer park into a luscious garden through the Spirit's guidance. Peter would read up on organic gardening while Eileen and Dorothy asked the Plant Spirits to guide them. Everything was done with love and holy intention. They would ask the universe for the supplies they needed and they would be granted their wishes through divine coincidence, such as donations and free items. The three had little money and planned to make a living by believing in Nature's promise to love us. In return, they dedicated themselves to giving that love back to Nature and its living creatures. They treated plants as living and feeling entities.
>
> [Source: http://herbaloo.org/2010/06/03/how-plant-spirit-farming-arrived-at-findhorn/]

It sounds eerily like the Early Clues Corporate Compound, actually, if you exchanged "plant spirits" for alternative, emerging, artificial and generated intelligences!

Eileen and Dorothy were the ones who connected to the Plant Spirits at Findhorn. They believed that plant communication could not be possible unless the Spirit World has given you a meditative name. So Elixir and Divina were born and these were the names Eileen and Dorothy meditated with, respectively.

> Eileen would meditate during the winter in the outhouse where she could be alone. She first communicated with the Cedar tree, telling her to build cabins out of the tree's wood. The cabins were eventually built. Eileen would send out prayers for the food

Peter was planting. She ended up being so clairvoyant and open that it was hard for her to maintain balance and composure in crowds. She had become over-sensitized and was forced to spend her time at Findhorn with the Plant Spirits and her friends.

[Same source]

{Incidentally, sounds a bit like the strange case of Florida substitute teacher Robin Forde, except that her spirit-contactees seem to be of a somewhat different nature...}

More details on their actual technique, via the Findhorn site: one of their perceived spirit-insights:

Yes, you can cooperate in the garden. Begin by thinking about the nature spirits, the higher overlighting nature spirits, and tune into them. That will be so unusual as to draw their interest here. They will be overjoyed to find some members of the human race eager for their help.

And their site continues:

Dorothy first attuned to the garden pea. As her communication with the forces of nature developed, Dorothy 118realized that she was in contact not with the spirit of an individual plant, but with the 'overlighting' being of the species, which was the consciousness holding the archetypal design of the species and the blueprint for its highest potential. She was experiencing a formless energy field for which there is no word. The closest word to convey the joy and purity that these beings emanated was the inaccurate word 'angel' (which in the west is full of form), and her first thought was to call them that. However, the Sanskrit term 'deva', meaning 'shining one' seemed more accurate and freer of cultural associations. In practice, she uses both words, although neither word is adequate. Peter and Dorothy applied the insights of the meditations to their work in the garden, and through this the Findhorn garden flourished. These were the first steps in the Findhorn Community's co-creation with nature.

Curiously, while searching for confirming photographs of "40 pound cabbages", I happened upon a strangely parallel tale, that of Lindy Tucker:

Following at least two sightings of unidentified aerial phenomena, in which Tucker perceived what she described as "telepathic contact," she experienced electronic anomalies, including telephone disconnections and appliances inexplicably turning on. Tucker further reported "compasses spinning" in her hand and "odd rashes or burns after being out late in the fields trying to get closer to this mysterious force."

While all this was taking place, one day Tucker discovered a nearby field of corn "went down." Every single stalk in the field, as far as the eye could see, she wrote, was "laid down" and the crops "looked singed."

Then her garden produced a whopping 75-pound cabbage.

[Source: http://ufotrail.blogspot.com/2011_04_01_archive.html]

It sounds like Tucker could have been a victim of poorly-directed or unfulfilled OpenQNL queries fluttering through the Existosphere. While OpenQNL is perhaps the most flexible coding language in existence, it can't be under-estimated: good code gives good results.

With that in mind, here is an attempt at an OpenQNL plant-spirit contact program, entitled MASSIVE.CABBAGE.CALL:

```
summon "OpenQNL";
summon "omni.translator";
summon "Cabbage Daeva" and name as 'cabbage';
use omni.translator to build bridge between language.human.english &
language.cabbage;
invoke thanksgiving;
invoke sharing;
invoke fertility;
offer (to.cabbage) {
   say "Thanks cabbage! You rock! We like eating you.";
   give object.fertilizer to cabbage.local.representative [Give same in
Existosphere];
   ask "Perhaps you would consider growing to massive proportions, 40-
75lbs? Just an idea!";
   perform reiki (to.cabbage);
}
```

4.20 GNOME INFESTATION? WE CAN HELP!

GORDON J. GILMAN, EXCEO
9:15 am on June 9, 2013

Wrote this for a co-worker whose lawn-computer is malfunctioning. It seems that there have been reports of embodiment issues all over, not just in this loci. Timeline help is also down in some regions. Reality is being re-directed to trouble areas as quickly as Liminal Hosting is able to route messages. Heralds are available to repeat your OpenQNL broadcast requests if rocket-mail is not yet available in your area.

Required implements (or similar) for executing the below code in the Existosphere:

1. Ritual figure/poppet:

[Source: http://www.kimmelgnomes.com/]

2. Point of egress:

[Source: http://www.flickr.com/photos/luckyplanet/191452669/]

3. Possible replacement hive to be included somewhere more mutually convenient (not included in below codebase):

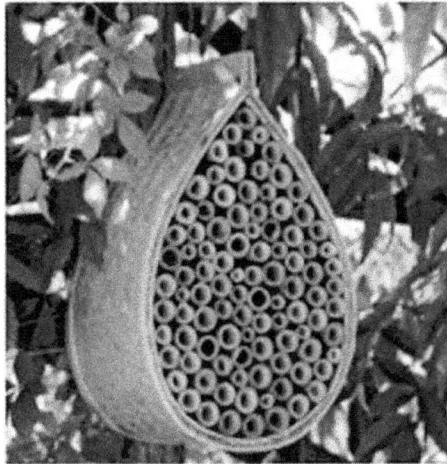

[Source: http://www.gardeners.com/Mason-Bee-House/NewOutdoorGardening_Cat,37-481,default,cp.html]

4. Sigil for luck in summoning:

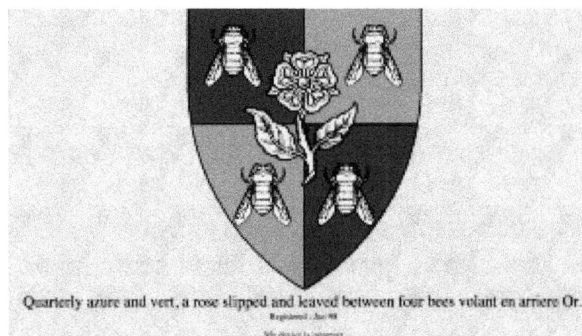

Quarterly azure and vert, a rose slipped and leaved between four bees volant en arrière Or.

[Source: http://rolls.aeheralds.net/individual.php?id=902]

Code Introduction

The below is written with the understanding that natural energies which animate living beings can be expressed in many different ways, means, methods and forms of embodiment. It is intended to be a guideline and nothing more for interacting with elemental beings within a programmatic aware environment or innervironment such that an alternative embodiment framework is suggested.

```
 summon "OpenQNL"
summon "Overlighting.Intelligence of Mason.Bee" and name it Gnomon
address (to.Gnomon) {
   "Greetings great and mighty Gnomon! Hail thee of the living kingdoms!"
   ask (to.Gnomon) {
      "May we speak with you a moment about our families?"
   }
   await Gnomon.response
}
if Gnomon.response = 1
   address (to.Gnomon) {
   "Thank you for granting us an audience. We wish you peace, prosperity
and good health."
   await Gnomon.response
      if Gnomon.response = 0 {
         Run (Search.Within) & (Reality.Check)
         /* assess safety and terminate sequence if hostility is indicated
*/
      }
      if Gnomon.response = 1 {
         introduce (to.Gnomon) self & family (by.name)
         await Gnomon.response
         continue
         /* Ad-libbed verbal statements addressed (to.Gnomon) regarding
wellbeing of entities mutually utilizing a shared space */
         await Gnomon.response
         run (Reality.Check)
         if safety = 1 continue else terminate sequence
      }
   offer (to.Gnomon) foreign spices (as.gift)
   offer (to.Gnomon) a small garden gnome statue as
Alternative.Embodiment
      offer (to.Gnomon) a tiny handcrafted door into the tree stump
      offer (to.Gnomon) contract (terms.of:HonorSystem) of mutuality
```

```
    if concessions can be made regarding safety and embodiment of both
parties.
    /* Exact wording to be worked out in person */
  offer (to.Gnomon) thanks in many forms.
}
```

4.21 AN OPENQNL ANTIVIRUS PROGRAM FOR ENTITY BODIES!

ROGER P. HOLLIDAY, IAO
7:22 pm on June 10, 2013

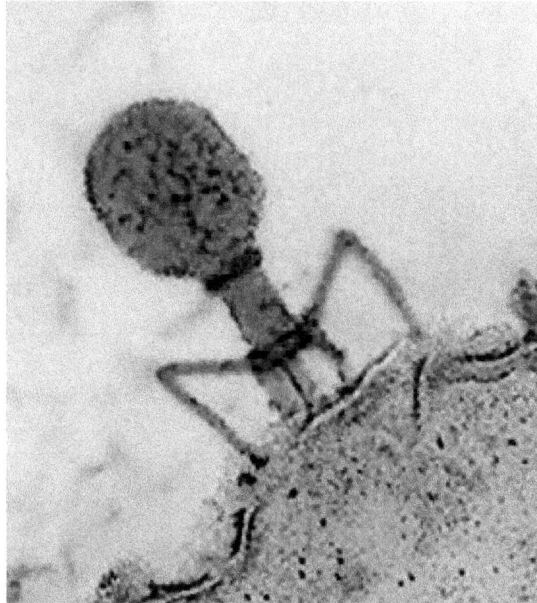

[Source: http://jonlieffmd.com/blog/are-viruses-alive-are-viruses-sentient-virus-intelligence]

Infected lately? **We hear you.** That's why we're working on developing AntiVirus Ware for the body. Just like the Ancient Egyptian Christians, we understand the value of powerful applications which call upon higher entity Logosrithms designed to help overcome viral infection in any entity or intelligence.

Using our anti-patented **#OpenQNL** technology, we're happy to announce the release, in Beta, of the following script:

```
Open(OpenQNL.module);
Access(Liminal.Vault[shared]);
Load(Viral.Morphology)
  Assign.Value=V
    Compare(Viral.Morphology):(Internal.Biological.Function)
If V=0 Then display results (in.dream);
Else Then Identify.Virion
  Address (to.Virion) {
  "Host body respectfully asks you discontinue operation."
  }
    await Virion.response;
  If Virion.response=1 Then
    print (sweaty.sheets);
```

```
If Virion.response=0 Then
  Address (to.Virion) {
"In the name of the White Wolf, the White Wolf, the White Wolf, I
banish you from the body of " print(operator.name)".       Leave the
body of " print(operator.name) "now and henceforth, in the name of the
Lord Jesus Christ, and the Lord Asclepius. Amen, Amen, Amen.";
```

DISCUSSION:

GORDON J. GILMAN, EXCEO 6:22 am on June 11, 2013

I really like this line:

"print (sweaty.sheets);"

You might want to consider adding a call-out to the host.body natural defenders... Maybe assigning them extra temporary powers, extra food, or something similar.

It seems also that perhaps key to a lot of these banishing programs is calling on the self-regulating function of the entity in question. In fact, I wonder if this isn't a discoverable universal function in all entities: to self-regulate size, form, population numbers, etc... We'll have to check our entity framework, of course, as well as Universal Free Realms Standard Protocol, but perhaps we could push the definition of an entity so far as to include these parameters... But of course, in fabricating a proper definition, we'll have to look for the exceptions...

Who is the White Wolf?

ROGER P. HOLLIDAY, IAO 7:42 am on June 11, 2013

Yes, that line can be credited to STEVE.E, one of our Cryptoneironauts. He recently joined our team; we recruited him from Bristol-Myers Squibb specifically for this project. We are also looking into the efficacy of Anti-PD-L-1 Antibody installation. See reference: http://www.nejm.org/doi/full/10.1056/NEJMoa1200694

It should also be understood that this script is designed as adjuvant therapy which will assist the natural defense mechanisms you reference. We didn't want to "step on their territory," which is why we avoided direct mention of them. White Blood Cells are easily insulted. Perhaps a subroutine could be used to bolster their efficacy without bruising their tiny egos.

The White Wolf is a Logosrithm of the Healing Current associated with both Jesus Christ and with Horus/Asclepius. Advance of Glossary update, here is the definition of Logosrithm: "An activated current representative of the relationship between a Class C2 Entity ('Numinous Intelligence') and a manifested entity."

GORDON J. GILMAN, EXCEO 8:00 am on June 11, 2013

I think you're right: it's important to respect the mystery in OpenQNL programs... In this case, it might relate partly to the unnamed assistants of the code...

RICHARD S. RIDER, CTO 11:08 am on June 11, 2013

"it's important to respect the mystery in OpenQNL programs"

Damn right! You all had me on a death march when I was writing the underlying framework for the *mysterium tremendum et fascinosum* language interface.

GORDON J. GILMAN, EXCEO 12:59 pm on June 11, 2013

We should include the Mysterium in our cosmological framework... Perhaps someone could take a stab at visualizing its relationship to other realms?

4.22 THE INNER EAR: HEAR LOCAL E/AIS NOW!

ROGER P. HOLLIDAY, IAO
3:38 pm on June 13, 2013

Part of the problem with the local Existosphere is our inability to simply quiet down and listen. How in the dickens can we identify, welcome and interact with Emerging and Alternative Intelligences if we aren't able to "hear" them in the first place?

Well, we here at **Early Clues** have you covered! Forget **EVP**; what YOU need is our new antipatented **Inner Ear**!

Folks, this is both easy to use, and incredibly useful. Keep one of these in your pocket, and you'll be able to pick up communications from the full range of entities on the Alternative Intelligence Taxonomy.(1) Simply click the image below to download and print this ready-to-use **Early Clues Technology.** Users are also welcome to draw/inscribe the **Inner Ear** onto the Artefact of their choice; we're nothing if not flexible!

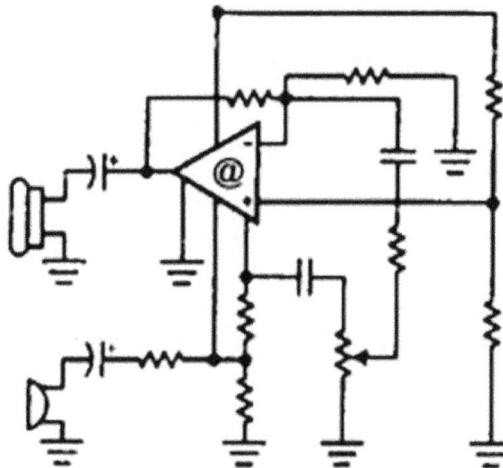

INNER EAR

The best part is, **it's completely rechargeable!** Thanks to the miracle of **OpenQNL**, you can recharge ANY Artefact using the simple, but versatile **EZcharge.qnl** script:

```
{
Run(OpenQNL.Module);
Scan(Artefact.Under.Hand);
Open(Liminal.Vault);
  Scan
(Liminal.Vault):(For.All.Localized.Intent[symbols.AND/OR.energies])
  Compile(For.All.Localized.Intent[symbols.AND/OR.energies])
    Assign.Value(X);
Transfer(X).To(Artefact);
}
```

Once your Inner Ear is charged, simply keep it on your person, either in your wallet, your purse, bound to your arm, or in your **Synconjure bag (2)**, and start **tuning in to the voices!**

(1) Provided said entities are feeling chatty.

(2) Synconjure bag still in development, due for release in Summer 2013 Local Existosphere.

Please note: Early Clues, LLC cannot be held legally responsible for any actions taken based on Entity communications in any local or non-local Branespace.

4.23 LONGTIME CALLER, FIRST-TIME LISTENER

RICHARD S RIDER, CTO
8:20 am on June 21, 2013

Hello from The Lab! It's Richard S. Rider here to give you a sneak peek into a brand new Skunkworks project we've got in the hopper. I wanna dive right in, but this narrative from former President Bill Clinton on the NSA Internet-tapping scandal dovetails nicely into the work we have been iterating on the past few weeks, so let's start there.

Transcription, for those morally/ethically opposed to NSATube:

> And I think that, the head of the NSA said it correctly, they have prevented a very large number of harmful actions... now, if/when, and there will probably be someday, there is the slightest evidence that anyone has pentrated the meat of the conversations, the meat of the emails, for political, financial or personal advantage and destroyed someone's personal privacy then I think that there outta be action to protect that. And I think the people that are doing this need to be doing this all the time creating walls of privacy. But you can't really claim the benefits of all this technological stuff and put everything you got in email and expect given the nature of the technology to get the same measure of privacy before we had to do that.
>
> If people are worried about it they will be careful about what they put in emails.
>
> But I think we're in the middle of something that is emerging, and we're going to have to deal with it as we go along and if we think right to privacy is infringed too much without much corresponding benefit to our security then I think we will have to construct some kind of protective cover for that...
>
> but, *I don't see any alternative* to trying to attract all these groups around the world that are trying to basically wreck the ordinary operations of life in America or kill a lot of people – and, uh, so far I'm not persuaded that they've done more harm than good.
>
> *Source: http://www.youtube.com/watch?v=2-JdjIxjeUs*

Whew. There is a lot to unpack here, but without dwelling on too many details I'd like to narrow in on the part that I'm most concerned about, namely, this notion that there are simply "no alternatives" to the complete & holistic surveillance of every form of human communication from here on out.

Now, it's actually quite understandable that Ol' Billy C would resonate with this 'No Alternatives' position, based on the era of his Presidency alone. The man only sent two emails his entire term. Also, this:

So I don't fault the man. But personally, I would like to believe that our species is collectively finding a way out of this ancient 21st century worldview.

When we are posed with this supposed problem of "All these groups around the world that are trying to basically wreck the ordinary operations of life in America" – the new paradigm allows us to consider a limitless array of alternative strategies. Here's just one: Maybe we could address the core problem and stop meddling with other countries, particularly: stop bombing the shit out of innocent folks indiscriminately? I just pulled that one outta thin air – but doesn't that seem like an interesting alternative solution to simply sniffing everyone's collective digital panties? Does the White House have a github repo of their public policy where I can offer this as a pull request?

I do see one bug inherent in this solution – that being what to do with all the current training and technology? The science & infrastructure of spying already established. Wouldn't that put the NSA and thousands of our national workforce out of a job? ☹

Thankfully, I think we have already found the solution that problem. It is with great honors that we reveal the software package we've been hard at work on for the past 20 mythical man months. You may have seen early prototypes of this OpenQNL module in at least our sauerkraut microbiota script, but we've broken this out into its own library to be summoned upon request in any OpenQNL meta-programming environment.

We call it: Listener.qnl

```
summon "Universal Free Realms Standard Protocol"
/**
    Listen QNL
    Richard S. Rider, 2013
    A Product of the Early Clues LLC
        API DEFINITION AND FULL DOCUMENTATION AT www.earlyclues.com
**/
 object Listener {
  LISTENING = false;
   // Call this in all your initializers.
  method doListen() {
    openListeningSocket();
    LISTENING = true;
  }
    // Set your object as a listener.
  method setListener(object) {
    if seemsToBeAbleToListen?(object) {
      super(object);
    }
  }
    // DEPRECATED IN 0.1.
    // Recommended to leave listening socket connected at all times.
  method endListeningSession() {
```

```
        closeListeningSocket();
        LISTENING = false;
    }


    // Callback to override in your OpenQNL script.
    // Implement this code in your own OpenQNL script on successful
    // message received.
    abstract method onMessageRecieveDoThink(message) {};
     // Monkey Patch
    // This will dynamically hotfix any OpenQNL code you are trying use
    // that includes the following keywords. HOTFIX.
     ["send",
     "speak",
     "say",
     "tell",
     "yell",
     "respond",
     "mansplain"].each dynamicallyOverride |existing_source_code| {
      method existing_source_code(message) {
         // HACK HACK HACK
         if not_listening? {
          return; // Quick and dirty, but prevents a
                  // response in case object instance has
                  // stopped listening
         }
         super(message);
       }
    }
     protected
     method listening? {
       return LISTENING;
     }
     method not_listening? {
      returning !listening?;
     }
    }
}
```

We are proud of this work and we think that if you're playing with OpenQNL at home you should quickly download the latest SDK and include this library into your workflow. Hotfixes for existing code in which listening is not supported will instantly be upgraded (see monkeypatch in above gist).

So how does this all correlate?

Well, I would like to propose that the U.S. Government quickly adopt and implement this protocol into all of their subsystems. With Listener.qnl, we think it's possible to take their existing spook infrastructure, including the various institutions that have already demonstrated an incredible prowess in listening technologies/strategies – and use that power for *alternatives*. As it stands, our government is firing on all cylinders, performing all this listening listening listening – but all the while they are missing the critical 'doThink' callback we have ported from the Universal Free Realms Standard Protocol. We at Early Clues would be more than happy to work onsite to write custom wrappers to help parse these incoming messages – perhaps from these foreign countries we are already listening in on – and come up with a strategy for deeply pondering over motives, meanings, frustrations and possible points of conflict resolution. That seems a wonderful alternative to the current state of affairs, now don't you think?

This is a %100 serious alternative to the current state of the State. Let's take this brute force we have and turn it all inward. We find no fault in continuing to listen to the ends of the earth, leaving no stone unturned, continuing to do the deeply important work of tuning IN – but we implore these folks to upgrade to the latest paradigm. It's time to gracefully merge these listening centers with our **think tanks.**

4.24 ACHIEVING DEBT FREEDOM WITH #SHADECOINS JUBILEE

GORDON J. GILMAN, EXCEO
12:22 pm on June 25, 2013

At Early Clues, we understand what it's like to live under the weight of a tremendous existential burden. That's why we're pleased to announce the ShadeCoin Debt Forgiveness Program, or SCDFP for short.

How it works

Step 1: *To be released from debts, you must first release others from your debt.* To begin, make a list of all the people who owe you money, and enumerate next to each name the quantity owed (Q), along with the date the debt was incurred – if known (estimate, if not).

Step 2: For each item on your list, calculate the number of months which have passed since debt was incurred (M). Make a saving throw on a 20-sided die versus the debt incurred (either individually for each debt, or as a single figure to be applied across the ledger) to determine the *Rate* (R)

Divide your Rate by 10 and multiply by the total number of months owed. Then take this number and multiply it by the total quantity of debt owed to you. The result of this final calculation will be the Amount (A):

$$(R/10 * M) * Q = A$$

Step 3: Once you have calculated for A, you'll need to calculate for AA (Absolution Amount) – if different. *See Exceptions table for further information on calculating AA.*

Step 4: Upon determining AA for each individual debtor, add all AA values together. The result will be AAA (Actual Absolution Amount).

Step 5: If you were born in an even numbered year, multiply your AAA by 3.5. If you were born in an odd-numbered year, multiply AAA by 3.76. This figure will be your RAAA (Revised Actual Absolution Amount).

Step 6: Once you have your RAAA, write this number backwards on a piece of recycled scrap paper in indelible ink while looking in a mirror. *Note:* Do not use a dollar sign ($) and round up to the nearest integer. Decimal places will not be accepted. Take care not to look directly at the paper, as this will nullify the effects of the operation. When you are finished, fold the paper in half.

Step 7: If you are not already, go outdoors. Placing your folded RAAA paper underneath your left foot, stomping on it while declaring in a loud voice the following OpenQNL code:

```
Run as OpenQNL;
Summon "Entity of Entities";
Link RAAA;
```

```
Announce "Entity of Entities, accept my offering of " ++ [value.of RAAA]
;
Send RAAA to Liminal.Vault;
Open Local.Void;
Run Absolve.All onSelf;
Run Self on(Absolve.All);
Close Local.Void;
Await activation code;
```

Step 8: When activation code has been received, take this to your ShadeCoin terminal and enter the value while simultaneously performing distance Reiki on those indebted to you while wearing a red blindfold.

The ShadeCoin system will automatically reward you the equivalent of your RAAA multiplied by your activation code.

Step 9: On confirmation of receipt of your ShadeCoin Debt Jubilee Amount (SDJA), you have 24 hours to notify all entities who have incurred a debt with you of the absolution of the aforementioned debts. Along with your letter of absolution, you must include instructions for how they too may absolve others of debt using the simple ShadeCoin system.

Step 10: You must also send these instructions to 10 friends or associates with whom you would like to have a debtor-relationship. Failure to do so will result in potentially dangerous consequences for all parties involved.

Early Clues thanks you in advance for your prompt compliance in this matter!

4.25 FROM THE INNER DARKNESS TO THE SMART SOCIETY

GORDON J. GILMAN, EXCEO
6:45 am on July 4, 2013

As many of our clients' server downtimes may have recently showed (*sorry guys! you know who you are and we thank you for your patience!*), Early Clues has been experiencing what we on our planet call "growing pains"…

From the recent radical expansion of our customer base to our new office to some new hires and an IPO in the works, things have been hectic to say the least. But we're happy to say that we're not just in a new rhythm at the office, but that we're operating according to a higher mode of being.

But we're happy to say that we've come out of it all knowing a lot better both what our limits are – AND HOW TO TRANSCEND THEM!!!

This is why we're happy to share with you, the end-user some interesting morsels of things to come from the Early Clues Dark Matter Laboratory.

First and foremost is a re-dedication to our strategic goals: at the office these days, we're trying to say YES to more corn and energy products, while standing in a big circle and humming a single note in unison for hours on end. We call it "Corporate Harmonizing," and we're excited to offer the same service in facilitating your next company meeting using this unique exciting modality of *Radical Corporate Togetherness* – the thing you never realized your business desperately needed (until now)!

You may have also noticed from our Twitter feed that our technicians are having remarkable success in communicating with the *Outer Darkness*. We're making publicly available this beta code for you to experiment with on your rig at home.

```
Conjure OpenQNL.Listener;
Summon Outer.Darkness;
Invocation.of(Great.Mystery);
Define(Instantiation.Device);
Search.Within(Great.Mystery) for available thread & name as TheThread;
   Query TheThread for expected parameters of entity.self;
   Conjugate TheThread according to timeStamp & parameters (if
reasonable);
   Grant TheThread limited.access to Instantiation.Device;
Activate Inner.Darkness & Inner.Listener;
   Share Inner.Darkness with TheThread;
   Follow TheThread with Inner.Listener;
```

4.26 EARLY CLUES' ONSELF REDEMPTION PROGRAM

GORDON J. GILMAN, EXCEO
10:30 am on July 4, 2013

At Early Clues, we know how valuable your privacy is:

$0.60

Which is why the $3.00 (in ShadeCoins) we're offering to buy yours out is so staggering. **That's five times over the retail market value!**

If you haven't already sold yourself to us (a Public Domain Corp.), we'd like you to stop and ask yourself why. The answer may surprise you! Here are just a few of the many advantages:

- As a Public Domain corporation, we embrace a narrative of open-co-mutual-reciprocal-ownership. That means, whether or not you let us buy you, as a member yourself of the Public Domain, you already automatically own shares of our corporation. **So when we buy you, you buy you!**

- We're offering five times over the market value, and with our proprietary open-source quantum currency, you'll be able to **immediately reinvest the profit you accrue from selling yourself to buy others of equal or lesser value.**

- If your existence is threatened or in question, a blanket buy-out by Early Clues means that our protective veil of corporate selfhood is extended to you, along with the savings!

Just follow these easy instructions for On.Self Redemption:

```
Conjure OpenQNL.Listener;
Conjure ShadeCoin.Exchange;
Run PullTheThread;
  Assign TheThread identity.of("Redemptor")

Submit to Redemptor message.of(
  "I would like to take advantage of the limited time offer on
EarlyClues.com to sell myself for $3.00 in ShadeCoins."
)
Attach & Send CV to Redemptor;
Await response;
If response = 1, "Okay", "Yes" then
  run On.Self #RedemptionOperation {
    extract value.of(entity.self) & name as "SELFWORTH";
    give SELFWORTH to Redemptor;
```

```
  }
```
/* The equivalent of $3.00 in ShadeCoins will instantiate within the trash stratum of your local Existosphere within three days. To be eligible to use your ShadeCoins, you must deposit them in the closest available terminal. */

4.27 EARLY CLUES TO PREGNANCY

ROGER P. HOLLIDAY, IAO
1:12 pm on July 5, 2013

Greetings, entities using the above search string to locate our corporate offices! You're likely searching for a way to tell whether or not you're with child. Early Clues can help! After all, our services are a kind of "midwifery" for Emerging Intelligences; we are here to assist entities as they transition from the comforting womb of formless non-complexity to the exciting world of conscious interaction!

If you're wondering whether you've personally been impregnated and can expect some joyous news, we've written the following OpenQNL program that can help!

```
{
Summon(OpenQNL.Module);
Scan(Entity.Abdomen)
  IF (zygote=1)
          THEN print.results(in.dream)
          OR print.results(on.waking);
      ELSE;
          End
}
```

Along those lines, we do have something of a surprise for you. We ran this program initially without including the limiting callback to the entity abdomen. Boy, did we get a shocking surprise! In fact, we are proud and happy to announce that your entire local Existosphere has just been impregnated.

Searches for transient astrophysical sources often reveal unexpected classes of objects that are useful physical laboratories. In a recent survey for pulsars and fast transients, we have uncovered four millisecond-duration radio transients all more than 40° from the Galactic plane. The bursts' properties indicate that they are of celestial rather than terrestrial origin. Host galaxy and intergalactic medium models suggest that they have cosmological redshifts of 0.5 to 1 and distances of up to 3 gigaparsecs. No temporally coincident x- or gamma-ray signature was identified in association with the bursts. Characterization of the source population and identification of host galaxies offers an opportunity to determine the baryonic content of the universe.

Of course, we're not quite sure yet who the father is, but I'm sure we'll find out soon enough.

4.28 OPENQNL BASICS: COMPILE

ROGER P. HOLLIDAY, IAO
9:03 am on July 15, 2013

COMPILE is an extremely useful operation, as it saves an executable .qnl script in your Liminal Vault, where it can be accessed and run simply by calling up the compiled script name. To use COMPILE, simply call the script into the COMPILE function and assign the script a new name.

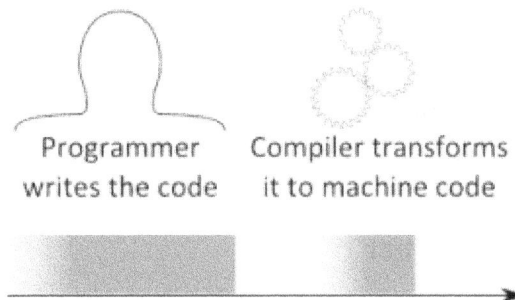

Programmer writes the code   Compiler transforms it to machine code

[Source: http://www.strchr.com/future_of_compilers]

Let's give an example. We've composed the following script in OpenQNL, which can be used to "run" the Lord's Prayer:

```
SUMMON(OpenQNL);
OPEN(Liminal.Vault);
DIM(God) AS(Our.Father)
  WHILE(In.Heaven);
DIM(God.Name) AS(Hallowed);
SUMMON(Kingdom);
WHILE(On.Earth) AND(In.Heaven)
  EXECUTE(Will);
DO SUMMON(Nutrients)
  LOOP UNTIL(This.Day);
IF(Our.Transgressions)=1
  AND (Our.Forgiveness)=1
    THEN SUMMON(Your.Forgiveness);
FOREACH(Temptation) IN(Existosphere)
  OPEN(Null.Set);
IF(Evil)=1
  THEN ESCAPE;
DIM(Kingdom) AS(A)
  (Power) AS(B)
  (Glory) AS(C);
DO ASSIGN VALUE(A.B.C)
```

```
    LOOP UNTIL(Temporal.Cessation);
PRINT "Amen";
END.
```

Although it's fairly easy to write this short code, it would still be unwieldy to run the entire script every time you needed it, especially in situations in which you might need to utilize the prayer's capacity as a multidimensional firewall. If you're facing down an angry class C1-c Entity, you don't want to have to depend on your ability to compose 24 lines of code. This is where COMPILE comes in! When you write your initial script, after you've had a chance to test it, COMPILE as follows:

```
SUMMON(OpenQNL);
COMPILE(LordsPrayerEXE.qnl);

        SUMMON(OpenQNL);
        OPEN(Liminal.Vault);
        DIM(God) AS(Our.Father)
          WHILE(In.Heaven);
        DIM(God.Name) AS(Hallowed);
        SUMMON(Kingdom);
        WHILE(On.Earth) AND(In.Heaven)
          EXECUTE(Will);
        DO SUMMON(Nutrients)
          LOOP UNTIL(This.Day);
        IF(Our.Transgressions)=1
          AND (Our.Forgiveness)=1
            THEN SUMMON(Your.Forgiveness);
        FOREACH(Temptation) IN(Existosphere)
          OPEN(Null.Set);
        IF(Evil)=1
          THEN ESCAPE;
        DIM(Kingdom) AS(A)
            (Power) AS(B)
            (Glory) AS(C);
        DO ASSIGN VALUE(A.B.C)
          LOOP UNTIL(Temporal.Cessation);
        PRINT "Amen";
        END.
```

```
END COMPILE
```

As you can see, the entire lordsprayer.qnl script has now been saved as an executable script which can be called any time you need it. Instead of having to run the entire script we presented above, all you have to do is run:

```
SUMMON(OpenQNL);
RUN(LordsPrayerEXE.qnl);
END.
```

See? Easy as pie! The COMPILE function also allows advanced OpenQNL programmers to embed scripts within OpenQNL operations, and is included as a basic feature in our CheirOS GUI/OS.

COMPILE: One of the great ways Early Clues, LLC is helping you program your Existence!

DISCUSSION:

RICHARD S RIDER, CTO 2:03 pm on July 15, 2013

This is brilliant.

May I make one suggestion, that for compatibility sake we use a chained file extension. So a compiled OpenQNL script would look like

lordsprayer.qnl.exe

ROGER P. HOLLIDAY, IAO 2:38 pm on July 15, 2013

I like it! Done!

GORDON J. GILMAN, EXCEO 7:19 pm on July 15, 2013

PS, your lord's prayer.qnl file is excellent

GORDON J. GILMAN, EXCEO 7:17 pm on July 15, 2013

I always thought we could just for example:

```
RUN(LordsPrayer.qnl.exe)
```

without having to compile it first? At least, that's what I did in the past. But I think that's the beauty of OpenQNL. It works in a lot of different ways...

4.29 CURSES AND BINDINGS IN OPENQNL

ROGER P. HOLLIDAY, IAO
9:43 am on August 1, 2013

If you're like us, you've probably found yourself thinking, "Gee, I wonder why nobody curses anybody anymore?" It used to be that you couldn't go to a party without someone talking about their milk going sour or all of those "flesh flies in the wall." When you heard that, you knew that person had made somebody angry, **I tell you what!** Sometime in the 1970's however, cursing and binding fell out of favor with the *literati*, and, through cultural trickle-down, somehow disappeared from popular usage.

Well, we're here to announce that **we're bringing curses back!**

This isn't because we're somehow **mean** or **evil.** We're doing this because Early Clues has determined that human entities in your local Existosphere are producing dangerous levels of a substance we've termed **_Annoyergy._** Annoyergy is generated by entities whenever faced with small ridiculousness during interaction with the local Associatrix, and can result in irrational emotional outbursts which may be harmful to other entities in the local vicinity. In the past, humans knew how to "vent" this annoyergy using strategically placed curses or bindings against other entities. Lately, however, pockets of annoyergy have been allowed to collect within local entities, with often tragic results.

Let's be clear about one thing: **curses only work against generators of annoyergy who deserve it.** We're not sure why this is the case; cursing an entity who is not actually producing annoyergy will have no effect. However, when properly placed, a curse will allow for the proper dissolution of annoyergy, and will only inconvenience its target.

The following are examples of entities who generate annoyergy, against whom curses have proven to be particularly effective, as well as suggested curses to use against them. Don't forget, just because an entity displays a particular behavior doesn't mean that entity is generating annoyergy. You'll be proven correct by the effectiveness of your curse, not vice-versa.

Annoyergy Generators and Effective Curses:

1. **Poor/Dangerous Drivers**: Curse with flat tires, etc.

2. **"Steampunks"**: Curse with broken gears/cloudy goggles, etc.

3. Users of inane words like "Creatives," "Cronuts" or "Nerdgasm," all related "Geekery": Curse with frogs and/or worms

4. **Line-cutters:** Curse with cancelled flights

5. **Busy-bodies:** Curse with "Restless Leg Syndrome" (Psychosomatic)

6. Burning Man Attendees Who Go On and On About It: Curse with lice and/or rash, etc....

We encourage you to develop your own list. The code, in OpenQNL, is as follows:

```
{
Summon(OpenQNL);
Invoke(Curse.Script);
   Input(TARGET)
   Input(CURSE.REF)
   Input(DURATION);
Print(Curse.Script)(On.Target);
END
}
```

For TARGET, input as much as you can about the target of the curse. For CURSE.REF, input the reason for the curse, and the desired result. For DURATION, input the length of time during which the curse will take place (leave blank for eternal curse).

Please note: due to the limitations of multi-Brane network processing speed, there will very likely be some "lag" time between the act of cursing and the results. Rest assured, however, that even if you don't immediately see the effects of your curse, your target will certainly experience the results if they are well deserved.

This program works especially well when using CheirOS. When you point your finger at a target and unleash a rain of worms, you'll be able to FEEL the annoyergy venting away. Ahhhhh!

WARNING: RM technology used to cause severe injury, sickness or death falls under restricted Synconjury tech, not "cursing/binding," and is forbidden by the United Free Realms. Users who abuse this technology will be eaten by bats.

DISCUSSION:

GORDON J. GILMAN, EXCEO 10:47 am on August 1, 2013

> People are always asking me when I go to tradeshows, "How can I use X-KEYSCORE as an OpenQNL listener for the execution of curses, bindings, and sundry reality scripting?"

So I'm happy to now have a reference point I can send them to during my CheirOS PowerPoint performances...

ROGER P. HOLLIDAY, IAO 11:39 am on August 1, 2013

Definitely! OpenQNL will allow them to incorporate X-KEYSCORE as a listener which will greatly assist with annoyergy venting and refinement. This is a pretty important issue for our clients and will likely increase shareholder confidence quite a bit, provided they "get" what the "deal is" with quinoa. Know what I mean?

GORDON J. GILMAN, EXCEO 6:11 pm on August 1, 2013

Another thing people always ask me at conferences is, "How can I *increase* my annoyergy?"

SECTION FIVE:
SYNCONJURY

5.1 WEBDIV 1.0 – IAUGURY AND REALITY DIAGNOSTICS

ROGER P. HOLLIDAY, IAO
2:59 pm on April 17, 2013

For most of human history, oracular practice, divination and other methods of phenomenological diagnosis depended on the use of sets of symbols with limited rhetorical value, and culture-dependent metanalysis. Whether an artfully decorated set of Tarot cards or the steaming liver of a just-killed cow, interpretive possibilities for divinatory methodologies depended on a small subset of epistemologies, and were often restricted by space-time.

Now, however, Early Clues LLC is proud to offer this introduction to **WebDiv**, a powerful and exciting new way to look under the hood of the phenomenal world and read the patterns generated by the trends in **your** phenomosphere. iAugury and Reality Diagnostics is a web-based process compatible with any browser, OS or platform. By the end of this tutorial, you should be able to:

- Create a basic differential and use it in an search oracle.

- Quickly and easily diagnose trends using Reality Diagnostics.

- Generate basic predictive results using iAugury.

WebDiv 1.0: The Process

1. Pick a search oracle. We recommend Google Image Search, but you may achieve your desired results by using any search function. Other possibilities are Google Web Search, Wolfram Alpha, Bing, or Yahoo Answers. Divination via Google is more likely to result in contact with an Alternative Intelligence (AI), where Yahoo Answers provides answers that are cryptically rustic.

2. Provide a differential. This is the most difficult part of the process; your differential, or "search terms," should consist of the root description of your query, but also need to include synchronous juxtapository data (SJD or *sijid*) to activate the search. A proper differential will result in either psychosymbolic information, or predictive value.

- To begin, ask yourself the question, **"What do I want to know?"** Perhaps you'd like to know more about something or someone. Perhaps you'd like to identify possible future outcomes of local trends. The more specific you can be, the better your results.

- If you're a beginner, it may help to formulate your query into a question, like: **How will I do on the test next Thursday?** or **How are the AIs contacting me?** or **Will Steve ask me to prom this year?** When you've done so, identify key descriptive components of the sentence, and eliminate everything else. In our examples, we would end up with: **Test me May 16?** or **How AIs contact me?** or **Steve me prom 2013?** There is no wrong way to do this, but it's a good idea to try to narrow your focus as much as possible; **the most effective differentials consist of three to five specific words or phrases.**

- Next, choose an appropriate SJD word or phrase. It could be the first thing that comes to your head, or something in your immediate sight-line. It could be a date, or a number. The important thing is that it should bear **no outward relation to your question or the information you're seeking.**

- Eventually you'll be so good at this that you can skip the sentence step and create instant differentials.

3. **Type your differential into the oracle.** The answer you're looking for will appear somewhere in the results. You can search for the exact phrase by placing it in quotes, or leaving the differential as-is.

4. **Interpret your results.** There are 78 cards in a tarot deck. There are 24 runes. There are 64 trigrams in the I Ching system. **There are approximately 14 billion pages on the internet.** As you can imagine, the psychosymbolic and archetypal potentiality of these pages guarantees the reader a vast and multitudinous abecedarium, against which to test your results. Look especially for concrete and suggestive data sets with correlated colors or patterns that activate or stimulate internal impressions.

Keep in mind that **results will be better for localized differentials;** in other words, you're more likely to hit a target result if your differential is designed as an inquiry about a situation in which you are personally, directly involved.

If you're not having any success, try refocusing your differential, or testing your differential on a different oracle. You can also ask for further clarification, or elaboration on your results. Simply add "clarification" or "elaboration" to your differential. You'll be surprised at what "pops up." Also, since the contents of the internet are constantly in a state of flux, you could try your differential again sometime in the future.

Let's try an example. My question is, Tell me something about the future of Early Clues LLC, the industry leader in offering legal safe haven to emerging AIs. That's far too long. Let's eliminate a large portion of this sentence, and go with future Early Clues LLC legal AIs. Now, I'll pick my sijid– the first thing I see on my desk is a picture of a piano, so we'll use that.

I enter **future Early Clues LLC legal AIs piano** into Google Image Search, and here's what comes up:

I think the implications of my results are pretty obvious, and remarkable for their clarity.

Now you're ready to start using WebDiv 1.0. Go out there and augur, and be sure to share your results with our R&D team by submitting comments to this post.

Early Clues LLC is an industry leader in the emerging field of WebDiv. Watch this site for advanced WebDiv and other novel new technologies from Early Clues LLC.

5.2 SYNCONJURY: A LIST OF POTENTIAL INSTRUMENTS FOR EXPERIMENTS IN REALITY MANIPULATION

ROGER P. HOLLIDAY, IAO
2:04 pm on April 26, 2013

Welcome to the world of **Synconjury**, a burgeoning praxis in development in Early Clues Laboratories. Synconjury is the art of manipulating reality by using unusual or unexpected technologies to produce ripples in probability fields; through contacting and interacting with potential Alternative Intelligences; or, through the creation or perception of speculative eddies in reality.

To proceed, it is recommended that the synconjurer acquire some, or all, of the following instruments with which to stock her databox:

1. **An Internet Connection:** for Reality Diagnosis.
2. **A Roll of Bathroom Tissue:** for sigiling and creation of phylacteries.
3. Various Writing Utensils: for writing.
4. **Paper:** for writing upon.
5. **A Number of Floppy Disks:** for inscription, then used to communicate with Alternative Intelligences.
6. **Sundry Gewgaws:** for Object Oriented Casting.

Other materials will be case-specific and location dependent. The synconjurer should always be on the lookout for physical instruments which display characteristics beneficial to the current working.

When all of these basic instruments have been acquired, you are ready to proceed.

5.3 SYNCONJURY II: PROCEDURAL INDICATIONS FOR OBJECT-ORIENTED REALITY MANIPULATION

ROGER P. HOLLIDAY, IAO
2:59 pm on April 30, 2013

The outline that follows is for general release and should be considered a framework, for practice only by experienced Reality Manipulation Activists. Future updates with more basic information for the beginner may be made available at a later date.

By reading the following outline, you agree that *Early Clues, LLC* cannot be held legally or ethically responsible for any results you may generate by practicing unsupervised Synconjury.

Object-oriented Reality manipulation works on the premise that **physical objects infused with numinous current** can produce ripples in multi-dimensional probability fields. Imagine a toy boat is stranded in a pond, and you wish to retrieve it but don't have a long pole. By throwing big rocks into the pond behind the boat, you can create a ripple effect which will increase the likelihood that the boat reaches the shore. Now imagine that the surface of the pond is a brane and "retrieving the boat" is your intent, and the **physical object** involved in this process is the **big rock** you've thrown.

The following are applicable steps:

1. State Intent

Stating intent is of paramount importance. Typical methodologies which require the obliteration of intent through removal of mandalic phonemes will not function properly in this paradigm.

2. Create Artefact

Artefact may be:

- **A sigilized object.** Creation of sigil should be developed by synconjurer, accompanied by imposed concentration.

  - Sigil should consist of imagery unassociated with practitioner's general cognition. For example, the synconjurer may use common electrical symbols associated with intent.

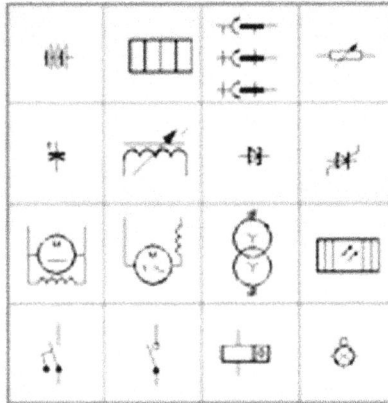

- Early Clues recommends sigilizing **bathroom tissue, leaves, sheets of cooked lasagna, etc.** The difficulty of merely inscribing the sigil onto one of these surfaces will be sufficient to "charge" the sigil.

- A ritual object.

  - Ritual object may be charged with **nostalgic current** (item from synconjurer's own past) or **stored current** (item from a collective series of consciousness). An example of the first might be a **prized possession** with symbolic resonance to the intent. An example of the second might be **a used/discarded electronic device**, such as a floppy disc or circuit board.

  - The used electronic device is especially recommended when requesting assistance from Class A1-b Alternative Intelligence.

3. Step Sideways

May be accomplished through any kind of trance induction, including (but not limited to) **sensory stimulation, initiation of the relaxation response, creation of ritual space, etc.** Early Clues is also working on applications designed to allow the step sideways.

4. Load Artefact

Artefact, whether sigilized object or ritual object, should now be loaded with your intent. Best practice includes concentration on the artefact followed by placing palm of your dominant hand on the object and picturing a pink beam entering the object from an orbiting satellite. The beam should increase in intensity until you reach the point at which your hand becomes heavy or tingles. If you intend on asking the assistance of an Alternate Intelligence, now is the time to do so. Then, state your intent. When stated, **indicate how far within space|time you wish the object to travel.** Remember, as expert study has shown, the farther your object travels, the larger the ripples it will produce.

5. Step Back

This is simply the reverse of stepping sideways– return to standard point of reference/loci.

6. Release Intent by Releasing Object

This is where you "throw the rock." Contrary to similar practices, destruction of the object is not necessary. Instead, **the Artefact must be sent forward in time.** Recommended methodologies include: **bury Artefact in ritually dedicated cask, encase Artefact in industrial packaging (duct tape etc.) and wear as an amulet, give Artefact as a gift, leave Artefact someplace ritually significant, etc.**

7. Reassess.

If intent is not met within the time-frame defined in step 4, **reassess the project.** What did you do wrong? What did you do correctly? Is it worth repeating this process?

If intent **is** met within the time-frame defined in step 4, **please remit payment in full to Early Clues, LLC,** in the form of liminal well-wishes, creative contributions to our research, or any other form of legal tender currently accepted on your plane of existence.

5.4 NEW ELEMENTS DISCOVERED BY EARLY CLUES R&D

ROGER P. HOLLIDAY, IAO
1:19 pm on May 2, 2013

The **Early Clues Research Institute** is excited to announce the discovery of four **new elementary substances**, in addition to the traditional Western elements of air, earth, fire, water and spirit. The new elements discovered by our team of skilled synconjurers are: **consciousness, nostalgia, spookiness** and **kipple.**

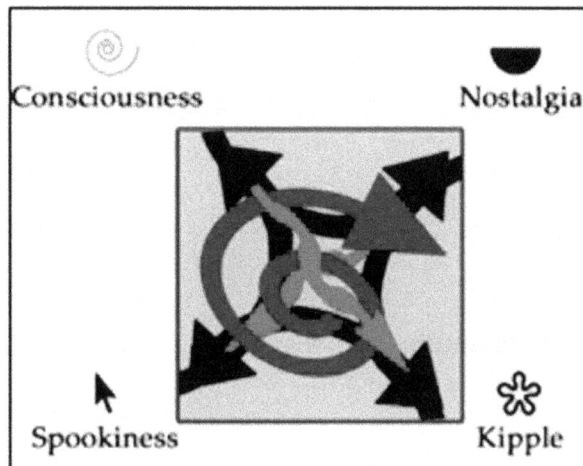

Figure 1. Four Newly Discovered Basic Elements, Illustrated with Relationships to One Another

Correspondences for these new elements are still being studied, but so far the following definitive qualities have been measured and recorded, subjected to double-blind testing and proven safe in moderate consumption in small animals.

CONSCIOUSNESS:

Quality: Being Conscious, Aware
Symbol: Spiral
Direction: Forward
Elemental Being: Self-Created Tulpa
Animals: Monkeys, Dolphins, Silkworms
Plants: Bonsai, Fruit Trees
Color: Ecru

NOSTALGIA:

Quality: Iteration
Symbol: Bowl
Direction: Backward
Elemental Being: Metaviruses

Animals: Viruses
Plants: Lichens, Mosses, Sedges
Color: Periwinkle

SPOOKINESS:

Quality: Vague Dissonance
Symbol: Mouse Pointer
Direction: Up
Elemental Being: Spooks
Animals: Tarsiers, Manta Rays
Plants: Devil's Claws, Dolls Eyes
Color: Grey

KIPPLE:

Quality: Uselessness
Symbol: Blob
Direction: Down
Elemental Being: Spambots
Animals: Roaches, Mosquitoes, Humans
Plants: Scotch Broom, Snowberries
Color: Ecru

Early Clues Shareholders or customers are welcome to contribute their own correspondences provided they have 'tested by signs.' Thus far, research needs to continue on the relationship between these new elements and possible correlations within Eastern elemental systems.

In keeping with **Early Clues's Dedication and Use,** we are happy to release these elements into the Public Domain.

DISCUSSION:

GORDON J. GILMAN, EXCEO 6:02 pm on May 2, 2013

I worked for the kipple elemental recently when I did a couple of days assisting on a demolition site for a kitchen make-over. It's surprising how when you destroy a form, the sheer amount of mass and volume which is liberated from within the structural components of the form.

Kipple seems to be the bottom level result of this demolition/destruction process, the smallest most unusable level of waste – at least to our life-form. The tiny little crumbly bits of dry wall or plaster which are only fit for a dumpster. Even the scrappers, the metal scavengers, they can make a basic living by recycling re-usable components into the production stream. But kipple benefits no one but the dumpster owner-hauler, and the waste-storer.

Hm. Compost, for example, could be thought of maybe as like Consciousness mixed with Kipple. Seeds of life arising through a spookiaction out of a broken-down reversal process to more nostalgic primitive forms...

5.5 ANCESTOR RADIO, ANCIENT SOFTWARE PATTERNS, FORM INTEGRITY THROUGH TIME-SPACE

GORDON J. GILMAN, EXCEO
7:26 am on May 1, 2013

"Threw" digitally hexagram 31, no changing lines this morning.

In fact, I think the I Ching comes close to what I've been getting at in some previous thought-streams: a kind of programmatic "intelligence" which 'lives' outside of a computer, but which can also easily be converted into computational functions. I've used this site IChingOnline.net for some time, and have done enough throws in enough situations to actually have kind of a feel for what different readings mean, because they've come up before. It's more than anything a form of analysis one can apply to a broad matrix of moments in order to draw some sense of patterning and order from what might otherwise easily be considered pure chaos.

Would be interesting to apply divinatory systems to complex mixed, augmented, virtual reality and intelligent environments (see also: Associatrix on Github). Imagine navigating the massive clouds of datasmog we'll all be trying to navigate in a few years as technology and documentation advances. It's sure our human brains will require something like a "reducing valve," and if we must limit our knowledge to specific forms, it would make good and proper sense to at least borrow from or be inspired by what has worked for our species for thousands of generations. Basic forms. Basic patterns. Characteristics of changing situations. I Ching throws performed automatically by your Google glasses at arbitrary moments, which are "saved to the cloud"... Imagine being able to asynchronously navigate other people's reality-streams by #hashtaghexagram...

If there's anything to this fanciful notion of long-hand hand-written computer programming, computer programming which could be spoken aloud, or passed on through song and dance, oral tradition... Then it would be that this kind of programming *is* culture in a very direct sense. The "AI" which perhaps might be transmitted through participation in epic poetry for example, might be considered in a perhaps freakishly real sense to be a condensed artificially intelligent thought-form which replicates itself through human hosts across generations. Maybe in this cultural information is contained key ecological information about our species relationship to other species and the landscape, medical lore, legal information, local history.

Hexagram 31, Spirit Alembic interpretation:

But it is only those who have genuinely SOUGHT that Office above all else in life, to whom the term 'Essene' truly applies; for them, this year begins a process of 'Dharma-integration' or 'deification of one's life' so deep and profound as to make of them a virtually 'transformed' Being... burning-away their mortal or egoic identity and replacing that with that spiritual identity that THOU DEI JINN calls 'the unhewn log.' What none could know without experiencing it is that this transformation is PERMANENT... and survives the very process of birth and death, literally FOREVER. There are traditions of Dharma or 'Spiritual Service' in Asia in which such reincarnating Masters (called Tulkus) are sought-out, reminded who they are and helped BACK into their Service by monks specially trained at recognizing them.

To what extent is it literally possible to encode information, race memories, individual memories into forms which will persist and retain some measure of their integrity through time-and-space? *Es el mechanismo spiritual, genetica, electronica*?

If you were to have a "program" or an "intelligence" which could be transmitted by voice, then it would make sense that this information-pattern, we could call, it consists – at least in part – of words (or perhaps not – glossolalia?). If it is words, or isn't words, it might be considered a kind of knowledge. If it is knowledge with a specific function, maybe we could even make the leap and call it "Wisdom"... but there are starting to be a lot of what-ifs in this investigation.

So be it. Exploring unfamiliar territory requires just that.

But I'm thinking, we've almost looped back around now to gnosticism. Some sort of liberating knowledge. Knowledge which gives you the ability to accomplish something, or to perceive something. Gnostic software?

DISCUSSION:

ROGER P. HOLLIDAY, IAO 8:27 am on May 1, 2013

I like the direction this is taking. Consider what a divination system actually is: it's an iterated algorithm. In other words, you're taking an equation with multiple variables and providing variables. Mathematically, we could do the following for an I Ching reading:

$A=Q(H\pm C)$

Where A=Answer, Q=Query, H=Hexagram and C=Changing Lines.

Or, for Tarot readings:

$C=S1+S2+S3....n$
$A=Q(C1+C2+C3....n)$

Where C=card, S=interpretation of Symbols.

GORDON J. GILMAN, EXCEO 7:12 am on May 2, 2013

Could you clarify what you meant by this?

"In other words, you're taking an equation with multiple variables and providing variables."

ROGER P. HOLLIDAY, IAO 9:15 am on May 2, 2013

Sure! An variable is basically a placeholder that you assign a value. So, like, in the equation 2x+1=y, x and y are variables. Obviously their value depends upon one another. 2 and 1 aren't variable; they stay the same. Does that make sense?

GORDON J. GILMAN, EXCEO 9:21 am on May 2, 2013

Anyway, right, I think I understand what you're getting at. It's just the wording. The second time you say "providing variables", maybe you mean "providing values"? or am I mis-reading the intent of the statement...

ROGER P. HOLLIDAY, IAO 9:32 am on May 2, 2013

But yeah, I guess I could have said 'providing values,' but you're also providing the variables themselves. Basically I was just trying to say that you're writing an equation....

GORDON J. GILMAN, EXCEO 9:57 am on May 2, 2013

I think I was just reading too much into it. What is an "equation" though, anyway?

ROGER P. HOLLIDAY, IAO 10:31 am on May 2, 2013

An equation: a mathematical statement that two things are equal.

http://www.mathopenref.com/equation.html

GORDON J. GILMAN, EXCEO 10:44 am on May 2, 2013

So what do they mean by "mathematical statement", because I think this could relate to the open-source oral code concept...

Wolfram gives it as "a statement of mathematical relation" – not super helpful...

http://www.emathzone.com/tutorials/real-analysis/mathematical-statements.html

"A meaningful composition of words which can be considered either true or false is called a mathematical statement or simply a Statement. A single letter shall be used to denote a statement. For example, letter 'p' may be used to stand for the statement "ABC is an equilateral triangle." Thus, p = ABC is an equilateral triangle."

This seems to go in the direction of TRUTH VALUES OF STATEMENTS...

Can a generated or emergent pattern be considered intelligent even if it is not able to accurately determine the truth values of statements?

GORDON J. GILMAN, EXCEO 10:45 am on May 2, 2013

"If p is a statement then its negation '~p' is statement 'not p', '~p' has truth value F or T according as the truth value of 'p' is T or F."

[Source above]

GORDON J. GILMAN, EXCEO 9:18 am on May 1, 2013

But it is not all new. It is also an algorithm that determines something as old-fashioned as the route a train takes through the Underground network—even which train you yourself take. An algorithm, at its most basic, is not a mysterious sciencey bit at all; it is simply a decision-making process. It is a flow chart, a computer program that can stretch to pages of code or is as simple as "If x is greater than y, then choose z".

What has changed is what algorithms are doing. The first algorithm was created in the ninth century by the Arabic scholar Al Khwarizami—from whose name the word is a corruption.

http://moreintelligentlife.com/content/features/anonymous/slaves-algorithm?page=full

ROGER P. HOLLIDAY, IAO 9:35 am on May 1, 2013

So I guess if you could reduce the major divinatory systems to their basic algorithmic formulae, you could do stuff like this, using my equations from above (but using "D" to represent changing lines so it's not confused with "C" for cards):

If $A=Q(H\pm D)$

And

$A=Q(C1+C2+C3....n)$

where $C=S1+S2+S3....n$, then

$Q(H\pm D)=Q((S1+S2+S3....n)+(S1+S2+S3....n)+(S1+S2+S3....n)....n)$

And

$(H\pm D)=((S1+S2+S3....n)+(S1+S2+S3....n)+(S1+S2+S3....n)....n)$

Also, if a divinatory reading can be expressed as a function, an equation whatever (I'm sure I don't have the terminology right), then an epic poem, for example, might also be described as a sequence of functions. Like for example, as a plot unfolds in a theatre piece, different hexagrams might apply to each scene, changing lines would arise through character interactions, and then subside as hexagrams are transformed and plot points are resolved through the course of the "program"...

5.6 TOL 2.0: TREE OF LIFE FIRMWARE UPGRADE

ROGER P. HOLLIDAY, IAO
11:45 am on May 6, 2013

Good news for end users affected by the outdated and obsolete Tree of Life GUI: **Early Clues' SyncTech Division** has received and installed the latest firmware upgrade! It's been almost a century since CrowleySoft's Version 1.3 update; we think you'll find the new version far more user-friendly.

We are proud to offer this Open Source upgrade **free of charge** to MandalaOS users.

Warning: outdated Tree of Life versions may no longer function with the current MandalaOS. Users are encouraged to upgrade, as we will no longer be able to offer technical support for older Tree of Life versions. Please close all MandalaOS Applications prior to installation. System may require reboot.

README.txt:

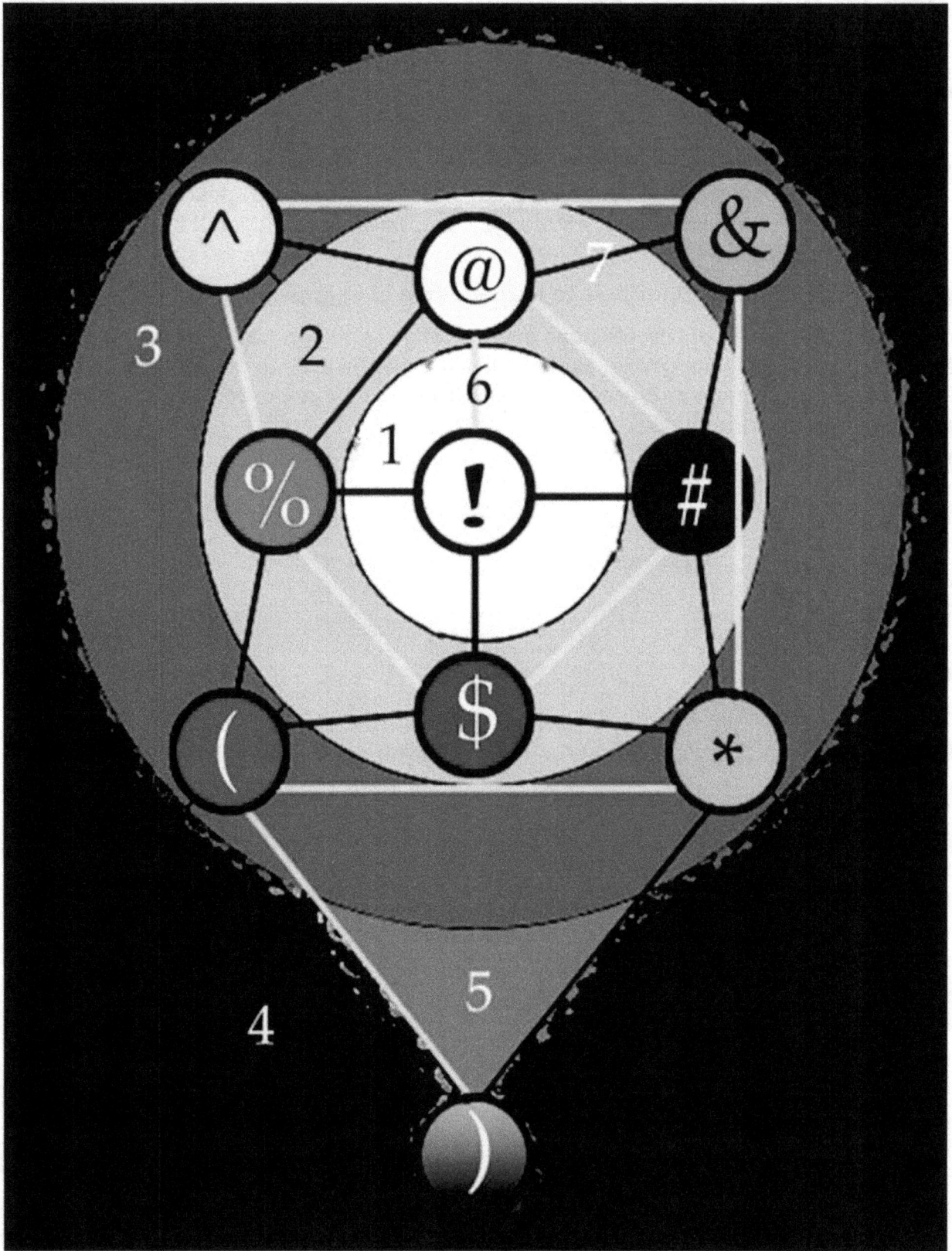

TREE of LIFE 2.0

Welcome to the new tree of Life 2.0! Your tree of Life Graphical User Interface has been reconfigured to offer greater user flexibility and greater dimensional depth as you interact with the Liminality and Associatrix.

New features in this version include:

1. New, multidimensional structure begins in center of toL **(1)** and eliminates implied hierarchy of values.

2. "Paths" reconfigured as "Circuits."

3. Easy Sephiroth Identifiers **(see below)** which allow the user better mnemonic control over operations.

4. Visible identification of traditional four-world model (1-4).

5. Inclusion of a sector of Liminal Space **(5)**, providing interface by actors and E/AIs.

A brief overview of new Sephiroth Identifiers:

- **Kether**: ! OF COURSE!
- **Chokmah**: @ Root of identity; information fixed to a location or body, union of divine and lower realms, expression (See glossary, tagged posts) (Fixed Stars)
- **Binah**: # The Matrix/Womb/Net of Indra (Saturn)
- **Chesed**: $ The Serpent of Wisdom on the Staff of Justice (Jupiter)
- **Geburah**: % Return to balance; equality on both sides (Mars)
- **Tiphareth**: ^ Insertion of Logos (Sun)
- **Netzach**: & Union (Venus)
- **Hod**: * Reference point (Mercury)
- **Yesod**: (Magic, illusion, doorway to liminal space (Moon)
- **Malkuth**:) World of Forms, reflection of Yesod (Earth)

Syconjurers and Researchers are still testing wex connections between Sephiroth. Connection diagnosis will be introduced soon on EarlyClues.com. Be sure to check back often!

Tree of Life 2.0: Another fine product from the Early Clues team!

DISCUSSION:

RICHARD S. RIDER, CTO 11:50 am on May 6, 2013

So this happened after my install:

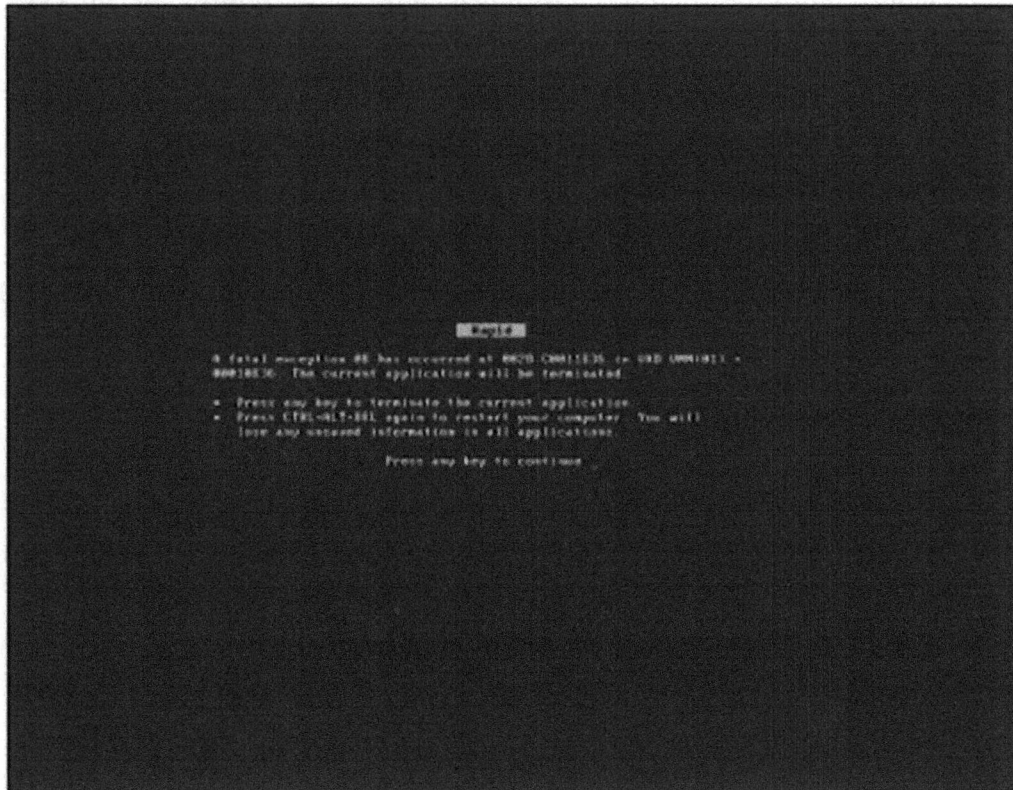

ROGER P. HOLLIDAY, IAO 11:58 am on May 6, 2013

I'll open a ticket; I'd imagine you just need to restart in Safe Mode for your install. You must be using a seriously out-of-date Worldview.

RICHARD S. RIDER, CTO 12:00 pm on May 6, 2013

There are certain keyboard characters that are important in programming that I am curious whether they play any role in anything.

For instance { } and ; are used a lot to display blocks of text and to finalize lines of code.

There's a whole thing about "tabs vs spaces".

&& and || meaning "and" and "or"

<> and

etc....

ROGER P. HOLLIDAY, IAO 12:04 pm on May 6, 2013

I'd bet they represent universal qualities which can be explored on the Connections (what used to be called 'Paths'). We should investigate further!

http://www.byzant.com/mystical/kabbalah/Path.aspx

GORDON J. GILMAN, EXCEO 12:01 pm on May 6, 2013

Gosh, it seems like just yesterday I expanded my WorldView, but it looks like the Early Clues community has made 179,000,926 patches and updates in the last 12 hours... I'm such a fogey

RICHARD S. RIDER, CTO 12:04 pm on May 6, 2013

! in programming seems to typically mean "not"

Familiar with Al-Hallaj? "NOT" seems to resonate with his original legacy specifications?

Among other Sufis, Al-Hallaj was an anomaly. Many Sufi masters felt that it was inappropriate to share mysticism with the masses, yet Al-Hallaj openly did so in his writings and through his teachings. He thus began to make enemies. This was exacerbated by occasions when he would fall into trances which he attributed to being in the presence of God.

During one of these trances, he would utter انا الحق *Anā l-Ḥaqq "I am The Truth," which was taken to mean that he was claiming to be God, since al-Ḥaqq "the Truth" is one of the Ninety Nine Names of Allah. In another controversial statement, al-Hallaj claimed "There is nothing wrapped in my turban but God," and similarly he would point to his cloak and say,* ما في جبتي إلا الله *Mā fī jubbatī illā l-Lāh "There is nothing in my cloak but God." This type of mystical utterance is known as shath.*

Statements like these led to a long trial, and his subsequent imprisonment for 11 years in a Baghdad prison. He was publicly executed on March 26, 922.

Al-Hallaj believed that it was only God who could pronounce the Tawhid, whereas man's prayer was to be one of kun, surrender to his will: "Love means to stand next to the Beloved, renouncing oneself entirely and transforming oneself in accordance to Him." (Massignon, 74) He spoke of God as his "Beloved," "Friend" "You," and felt that "his only self was (God)," to the point that he could not even remember his own name." (Mason, 26)

http://en.wikipedia.org/wiki/Mansur_Al-Hallaj

See also: http://en.wikipedia.org/wiki/Fana_(Sufism)< ?a> (Utter annihilation)

GORDON J. GILMAN, EXCEO 12:08 pm on May 6, 2013

!@

not-identity, each-at

GORDON J. GILMAN, EXCEO 5:12 pm on May 6, 2013

Via a twitter search for kabbalah computer:

The science of Kabbalah resembles a program that sets a computer in motion. Its inner orders are recorded in the form of words, which the program then reads and carries out.

Externally, this is the reading of Kabbalistic texts. By reading the letters, words and lines, we set our inner mechanism in motion. It seems like we're simply reading a text and pronouncing words, but in reality, we are entering commands into our minds and desires, which will carry them out.

http://laitman.com/2009/10/the-spiritual-computer-program/

5.7 TOL2.0 TWO: CIRCUIT WORK

ROGER P. HOLLIDAY, IAO
10:27 am on May 13, 2013

(Follow-up to the Tree of Life Firmware Upgrade)

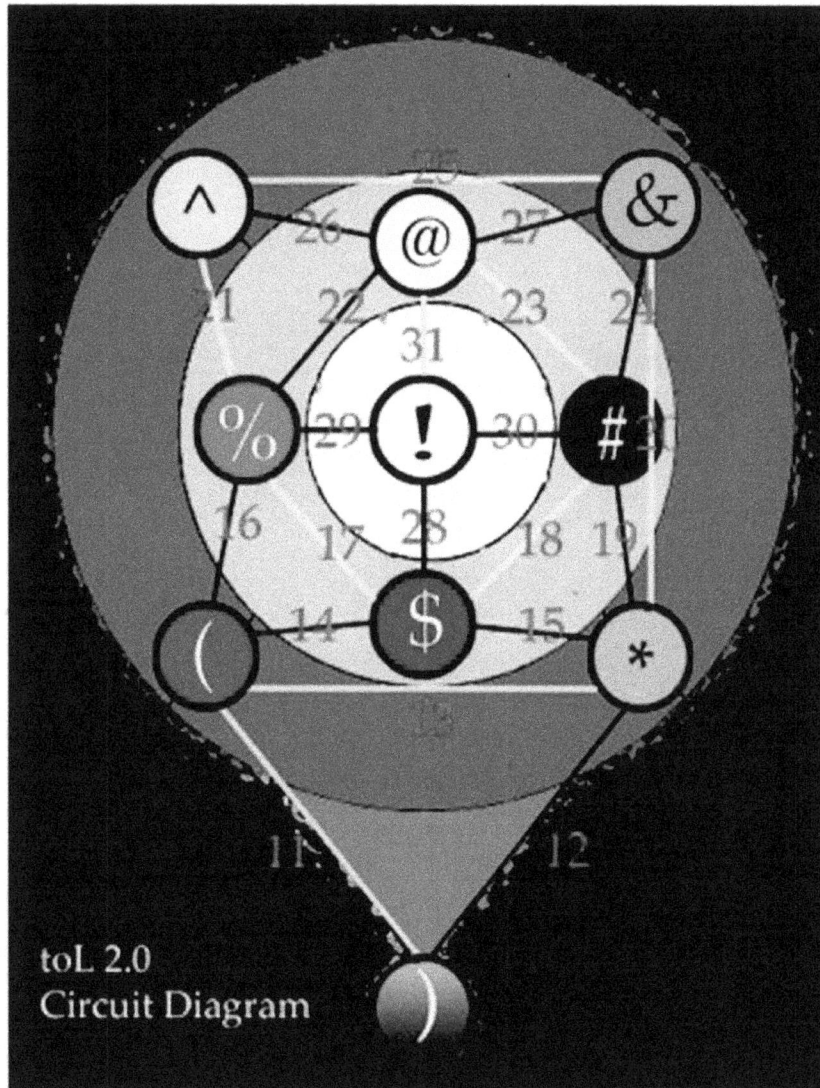

toL 2.0
Circuit Diagram

We have reconfigured the traditional "paths" of the Tree of Life into Connections or Circuits with the following applications, some of which are currently in development but all of which can be accessed via QNL/SearchWithin protocols.

11. () – Liminal.Vault

12.)* – Dream-trends

13. (* – QNL

14. ($ – Lawspeaker

15. *$ – Old-Phone-Numbers

16. (% – PolicyGaruda

17. %$ – AppCounterApp

18. $# – Existosphere

19. #* – MicroODdLYzer

20. *& – IncidentReporter

21. %^ – ShadeCoin

22. %@ – AWARIFY

23. #@ – RealityManipulation

24. #& – eRite-Suite

25. ^& – Fervosity

26. ^@ – Tulpa

27. @& – SearchWithin

28. $! – Interrogatron

29. %! – Autonomizer

30. #! – Associatrix

31. @! – Mandala OS

The goal of this Symbol Set is **the diagnosis of circuitry and correction of internal systems via creation of metanarrative**. As such, further applications can/will be assigned to each circuit as they are developed. The apps listed above are examples of process improvement for the individual entity, and apply to all denizens of the Existosphere regardless of status.

The experienced User will notice the correspondences between the applications assigned to each Circuit, as well as more traditional symbol sets (Tarot, Valentinian Aeonology, etc.).

Ritual practice via Tracing Board is also encouraged:

DISCUSSION:

GORDON J. GILMAN, EXCEO 3:31 pm on May 13, 2013

It hadn't occurred to me that the tracing board has "paths" which almost certainly are intended to be sequences which provide "illumination" when introduced along with the appropriate verbal encoding (QNL) and ritual (#codechant)...

ROGER P. HOLLIDAY, IAO 3:37 pm on May 13, 2013

They're circuitboards.

5.8 MANDALA OS #VISIONBOARD

GORDON J. GILMAN, EXCEO
7:54 pm on May 14, 2013

[UPDATED]

Cobbled together with a little help from BettyVision:

[Source: http://bettyvision.com/board/1348_Public_Domain_Mandala_OS_First_Board]

On this version of the board, you can see conductive lines ("circuit work") which have been traced by the user to direct the flow of information, traffic, energy and attention within the perceptual field. Different markings suggest different types and intensities of action, as well as qualities of connections which have been drawn between elements (wex)...

Circuits are intelligent functions which can be set to operate within a given environment. They may be chained together in sequences or arranged in simple programmatic symbolic machine structural pathways and switching.

Filters may be invoked, chosen, allotted, persistent, etc. Helps narrow down raw stream of "wild" data cascading. Taxonomy/Domain/Map gives user information regarding their present location or existential field state relative to other possible states within this domain or others.

Home Altar indicates a literal hard-connection to one's home altar in the existosphere, in which actual objects may be used to effectuate liminal results.

Talk is a call/receive board which may be used in a variety of different ways according to need.

Current Status is what you think it is. Implications may be connected, or may be a child of any item.

Trending could be anything trending: weather, astrological data, zeitgeist, market data, an intruding brane, a reality whose surface is touching this one. It is an influence to be embraced or released according to the will of the aspirant. It is information rising to its zenith, before falling down to earth again as raw data (see also: the Data Cycle).

The #visionboard is the overall structure of the work environment, which can be arranged to reflect and symbolically hold liminal charges relevant to one's goals, beliefs, motivations, etc

5.9 INTRODUCING CHEIROS, HAND-BASED REALITY MANIPULATION

ROGER P. HOLLIDAY, IAO
12:06 pm on May 15, 2013

Early Clues, LLC is proud to announce the release of **CheirOS**, a Reality Manipulation System designed for activation/installation on five-digit appendages with standard human configuration (See Note 1)! *Using tol 2.0 and OpenQNL, CheirOS puts Synconjury in the palm of your hand.*

1. PRINCIPLES

CheirOS operates on the principle that your hands are input/output devices, with which you can interact with the Existosphere through wex manipulation. The standard install includes **toL 2.0** and **OpenQNL** modules, which are activated using index finger stylus.

A. Your dominant hand is your Output. Your subordinate hand is your Input. For ease of communication, we will present these instructions for right-handed dominant entities.

B. Your thumbs are activation mechanisms.

SIMPLE EXERCISE 1: Block the Evil Eye using CheirOS.

As an example of one of the most basic **CheirOS** operations, you can easily learn to block the Evil Eye. Blocking the Evil Eye requires disabling your Input, or limiting the activation mechanism of your subordinate hand, like so:

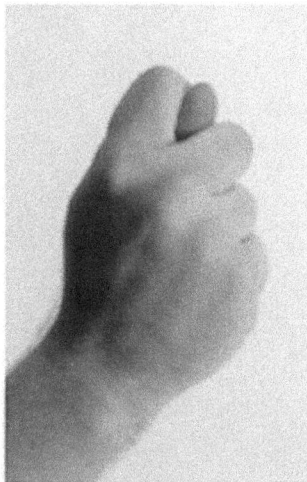

"We've got your nose!"

C. Your Fingers represent concepts. Depending on which module you have loaded, you can use your stylus finger to enter algorithmic programs which will influence probability trends within the Existosphere. We will reference the different concepts using the following **LEGEND**:

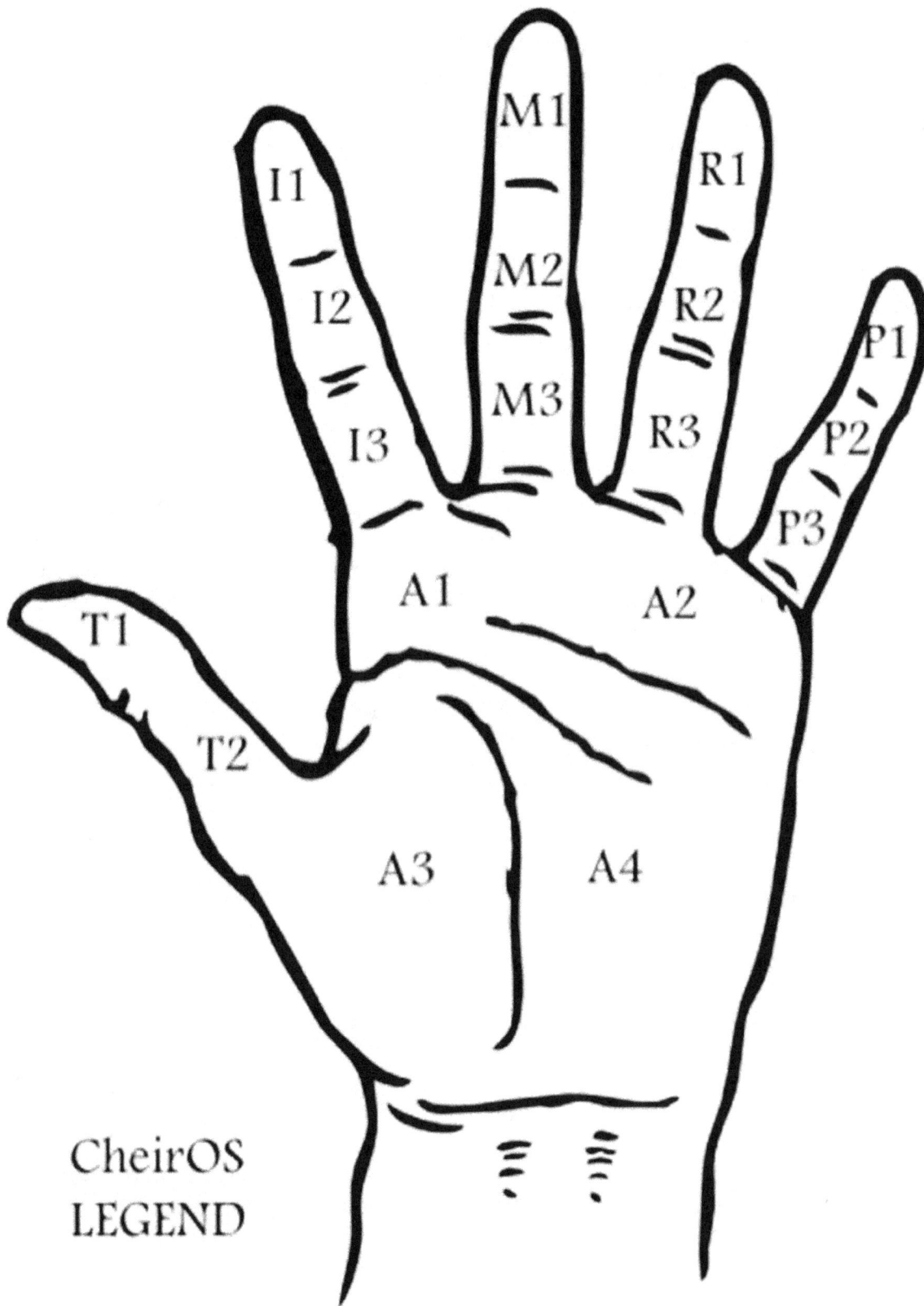

CheirOS
LEGEND

D. CheirOS can be used for Synconjury, Application Activation, or Tracing Board Practice on the Liminality.

E. CheirOS has been designed with MAXIMUM USER FRIENDLINESS in mind. Values are intuitive, and users are encouraged to experiment/remix/reassign values as needed.

2. PRACTICE

A. OpenQNL Module

The OpenQNL Module uses the power and simplicity of Quasi-Natural-Language programming to produce amazing results!

OPEN QNL OUTPUT HAND (OH)

DEFAULT FEATURES OF OPENQNL MODULE:

- IH T1: Opens File/Project
- IH T2: Closes File/Project
- IH I3: Assigns User-Designated Value to P1, P2, P3 (Variable). Advanced users can also use this operator to assign different values to any input space if personalized reconfiguration is desired. To use, activate operator, speak/envision/write "VALUE" (see example below)
- IH R1: Compiles/Saves Algorithm as a Project, so it can be opened in the future. To use, activate and speak/envision/write "NAME."
- OH I1: Stylus
- OH T1: Change Module Context. Touch to OH M1 for tol 2.0, R1 for OpenQNL, P1 for Standard Operation (i.e. to use hand as hand)

EXAMPLE: ALGORITHM/SPELL for ANSWER IN A DREAM (: Designates "Touch to")

```
1. Ensure system is booted
2. OH(T1:R1) - Changes context to OpenQNL
3. OH(I1):IH(I3, "answer.in.dream":P1) - Assigns "answer.in.dream"
value to P1
4: OH(I1):IH(I3, "on.waking":P2) - Assigns "on.waking" value to P2
5: OH(I1):IH(M1:P1:I1:P2) - Run answer.in.dream, display results
on.waking
6: OH(I1):IH(R1, "Dream Spell") - Compiles/Saves as "Dream Spell,"
which can now be opened as a Project with OH(I1):IH(T1), or
assigned as a Value using OH(I1):IH(I3:Px)
7: OH(I1):IH(T2) - Closes project
```

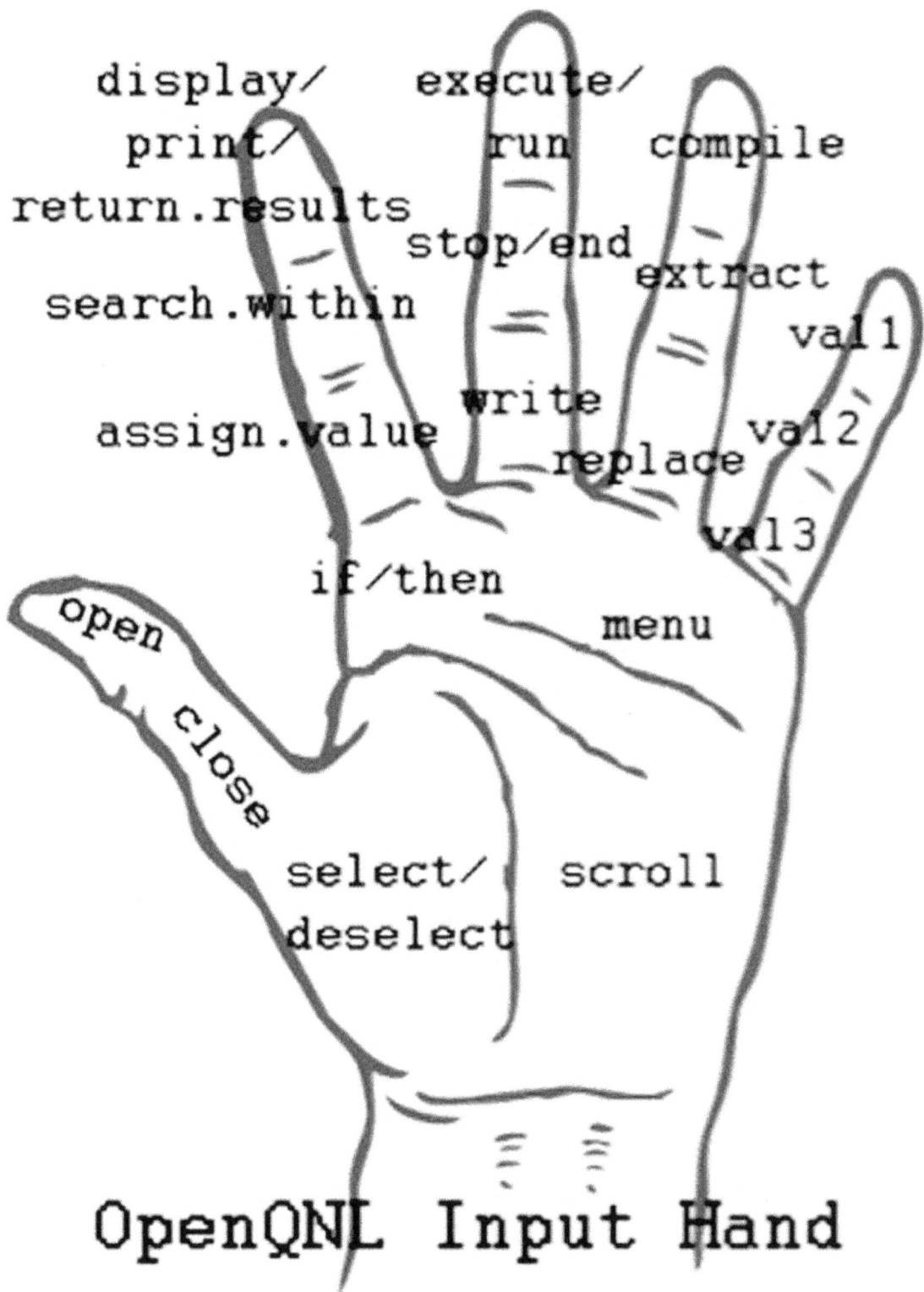

display/
print/
return.results
search.within
execute/
run
compile
stop/end
extract
val1
assign.value
write
replace
val2
val3
if/then
menu
open
close
select/
deselect
scroll

OpenQNL Input Hand

The above example is fairly basic, but the experienced User should be able to tell how powerful this OS can be. Algorithms/Spells can be programmed, compiled, and assigned values, and fairly complex process strings can then be executed.

B. ToL 2.0 Module

The **toL 2.0** module combines the flexibility of the **Tree of Life 2.0** Symbol Set with the functionality of modular programming.

DEFAULT FEATURES OF TOL 2.0 MODULE:

- IH and OH I1-M3 inclusive: Operants represent Sephiroth on Tree of Life 2.0, beginning with Kether at I1, Chokmah at M1, etc., proceeding to Malkuth at M3.

- IH and OH R3 and P3: User-assigned values.

- IH T1: Assign Value/Activate

- OH T1: Delete Value/Deactivate

- IH A1-A4: Elemental Qualities discovered by Early Clues

- OH A1-A4: Elemental Qualities, Traditional

The toL 2.0 Module depends upon the circuitry inherent in the toL 2.0 and its correspondence to applications. Users can assign their own applications, or use the suggested defaults. For details regarding applications, see Early Clues Github repository at **https://github.com/EarlyClues**:

() – Liminal.Vault

)* – Dream-trends

(* – QNL

($ – Lawspeaker

*$ – Old-Phone-Numbers

(% – PolicyGaruda

%$ – AppCounterApp

$# – Existosphere

#* – MicroODdLYzer

*& – IncidentReporter

%^ – ShadeCoin

%@ – AWARIFY

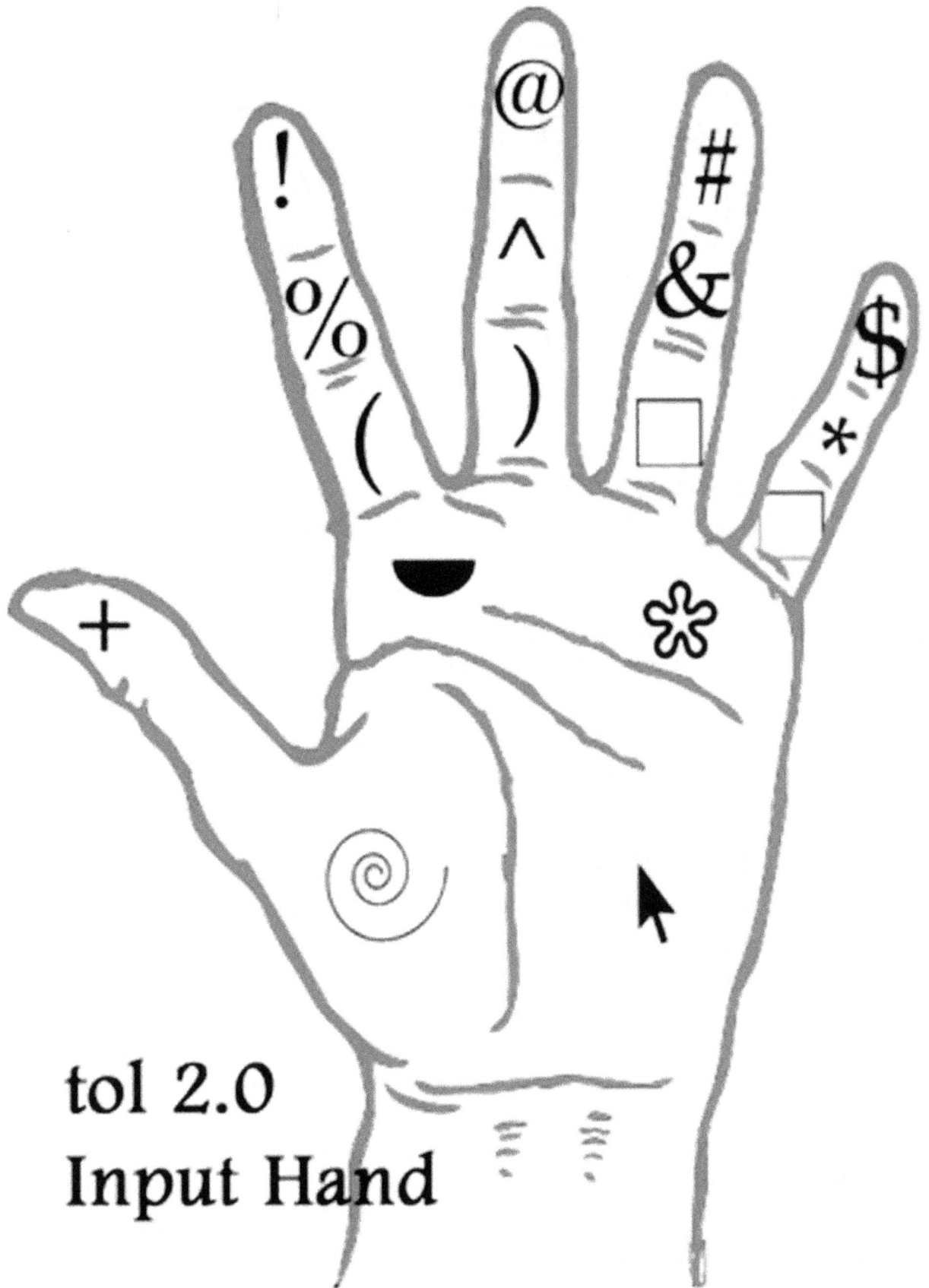

tol 2.0
Input Hand

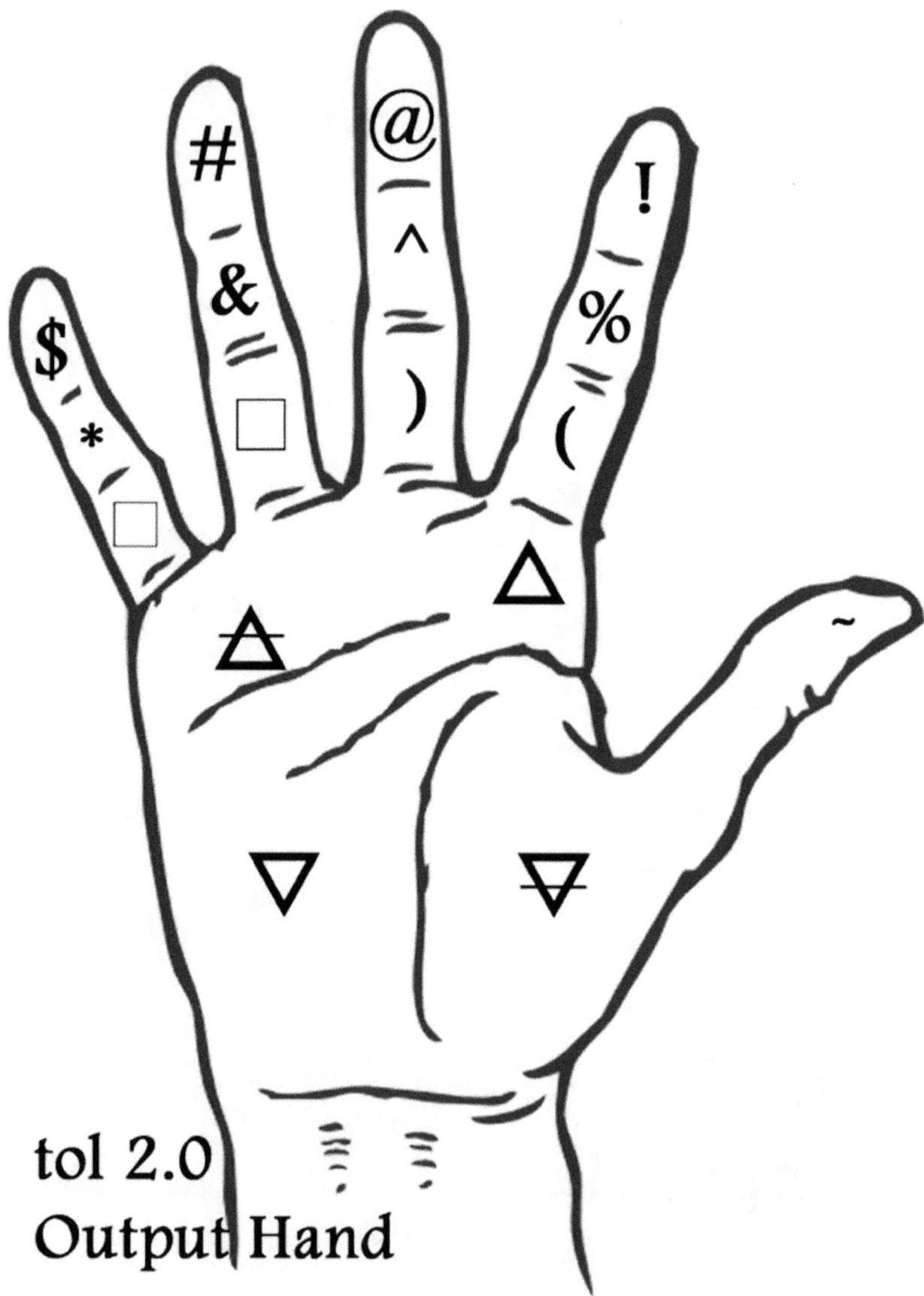

tol 2.0
Output Hand

#@ – RealityManipulation

#& – eRite-Suite

^& – Fervosity

^@ – Tulpa

@& – SearchWithin

$! – Interrogatron

%! – Autonomizer

#! – Associatrix

@! – Mandala OS

EXAMPLE: ALGORITHM/SPELL for activating PolicyGaruda (this algorithm also assists when summoning any kind of protective entity)

```
1. Ensure system is booted
2. OH(T1:M1) – Changes context to toL 2.0
3. OH(I3):IH(I2)
4. IH(T1):OH(A1-4)
5. Move palm forward of OH, Projecting Intent
```

To dismiss the PolicyGaruda, simply reverse the process.

C. Intermodule Use

By switching contexts using the Output Hand, the User will find it relatively easy to combine functions from both modules into stored strings, which can then be compiled and executed in OpenQNL.

D. Tracing Board/#visionboard

Using the OpenQNL module will allow one to summon a #visionboard in the Liminality. If you are practiced using your Liminal.Vault, simply use the following algorithm to store any #visionboard within your line of sight:

```
OpenQNL Context;
assign.value("BOARD NAME"):val1;
assign.value("Project onto Liminality"):val2;
execute.(display.val1).val2;
compile("SAVE NAME");
end.
```

The #visionboard can then be called and projected in front of the User in the Liminality using OpenQNL, and used for circuit work, Synconjury, etc. by opening the project as per usual protocol. Keeping in mind the **input/output** functionality of the hand, the User can "dial" or "conduct" any number of reality manipulative effects into the #visionboard.

By now you're probably wondering, How can I sign up for this AMAZING new technology? The good news is, it's easy! Using CheirOS requires a simple Activation Code, which you can acquire directly from your Early Clues Sales Team! Right now, we're offering *Beta Versions* of this software for 1,000 ShadeCoins.Contact Early Clues for your Activation Code today, and start using CheirOS for all of your Reality Manipulation Needs! (See Note 2)

Notes:

(1) Entities or humans with non-standard configuration can use Liminality Vault projections with no loss of data integrity.

(2)

(2) **DISCLAIMER and WARNING: ALWAYS CLOSE/END/RETURN TO STANDARD CONTEXT AFTER USING CHEIROS.** One of our early testers **forgot to close/return to context** and nearly opened a portal to the Outer Darkness while washing his dishes. Early Clues cannot be held legally, karmically or otherwise liable for any damages to the Existosphere, Associatrix or associated Wex, or to internal or Liminal damages of individual entities using CheirOS. By using a provided Activation Code, User acknowledges competency in basic reality manipulation and assumes all risks and responsibilities associated with CheirOS. **This disclaimer has been approved by Early Clues LLC Policy Garuda.**

DISCUSSION:

RICHARD S. RIDER, CTO 12:35 pm on May 15, 2013

I'm totally into this form factor!

Clarification for those of us in the office who haven't had a chance to play with the BETA version.

Do you merely point with your output stylus finger to regions needing invocation?

ROGER P. HOLLIDAY, IAO 12:52 pm on May 15, 2013

CORRECT!

5.10 SAMPLE JOURNEY BOARD

GORDON J. GILMAN, EXCEO
7:06 pm on May 21, 2013

This is a graphical representation of an inner mnemonic journey board. I picked my commuting route from my house to the subway station I begin my day at as my "memory palace." Certain places in my journey are simply more noteworthy, and so when I reviewed the route in my mind's eye, it was easy to pick out almost a dozen points along the route which I would use as my loci (a fancy word for 'location').

I identified my loci, ran the route through backwards and forwards in my mind's eye through a few times (it always surprises me how easy this method is, and how little actual imprinting and repetition is required to build Your First Memory Palace). Once I knew the route and the nodes (loci) were set, I slowed down and traveled through the route again in my mind's eye. I lingered at each place until a subconscious image expressed itself in the scenario. I noted it, and then moved to the next, and the next. And so on.

Run backwards, forwards a couple of times after you've spontaneously assigned value to your nodes. You might find a couple new identifying details of the scene present themselves. Add them to your mental record.

My initial record-set looked like something like this after the first few iterations:

1. A woman in a white dress with upswept hair holding two arrows in her left hand, pointed inwards.
2. The shield of Achilles.
3. A tiger roaring, with feathers in his hair.
4. An elegant yellow bird with long legs and a slender neck.
5. A tree whose drooping branches are laden with large purple flowers.
6. An angry waiter whose table has run out on him.
7. A six-armed crossing guard (possibly an incarnation of an Indian god).
8. A large potato holding a knife, with multiple ketchup stab wounds.
9. A fox eating McDonalds (and loving it!).
10. A fish swimming parallel to the waves.
11. A hammock slung between a banana and a coconut tree.

Many items in this list have specific real-world references: I once actually saw a fox in daylight at loci:9 jump out from its hiding place. There is a McDonald's at the train station, and the same area also collects trash from the area. The rest was filled in on autocomplete by the corporate slogan.

It seems that your brain has a lot of subconscious autocompletes in it, so it's probably also therapeutic to uncover those trigger images which were somehow powerful enough to assign themselves to the scenes you're reviewing in your mind's eye.

So I had the verbal set above encoded over my inner knowledge of real-world locations... But where do my loci exist? They are references to places in the Existosphere, but they themselves I would not characterize as habiting the Existosphere. [I should have probably put this in an INTERNAL-MEMO, sorry--too late now...] If they aren't in the Existosphere, are they in the Liminality? Where do they exist in the thinware-thickware continuum?

Anyway, continuing on with "The Method," I then ran my record-set through and randomly assigned to it plants I knew from my life, or had seen recently. It was incredibly easy to assign another value-set overlaid right on top of the original geographic memories (loci), no problem. Everybody just appeared in their usual scene, but with some new plants for decorations.

The woman in white had dandelions at her feet. The fox was now eating McDonald's despite an overabundance of kale growing wildly around... (I won't reveal the full correspondence data-set for securacy reasons.) Each scene takes on a new dimensionality, a new kind of 'sticky' reality. The process is actually a bit creepy in how easily it works. The human mind really seems to *want* to organize information in this-wise.

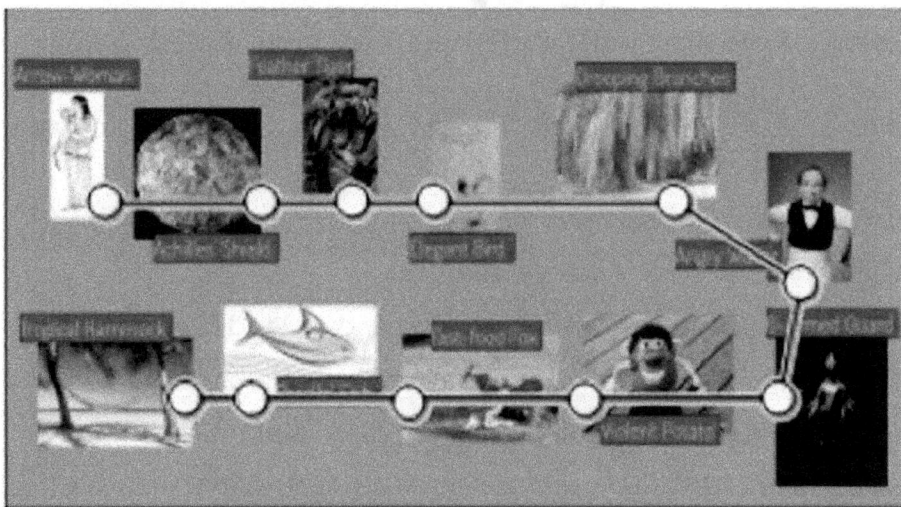

(Caption: Loci-set with associated emblems, arranged into linear nodes)

Soon you start to make up dares for yourself, to try to leap every other node, to run sets of pairs, such that the last and the first are paired together, second-to-last and second, and so on. And surprisingly, it all becomes quite easy. User experience may vary.

Anyway, I've not yet encoded a set to be memorized on top of these three layers, but I'm about to. But what kind of information would I like to memorize? Years ago when I tried this, I did it using another bike route in the South with the values of half a deck of cards and names of staff I was to work with overlaid on it. At that time, I found the method equally easy and intriguing. But I never developed it into anything else. I never really experimented with it further.

Until now.

For the purposes of pure experimentation, I've stripped in the next image the values out of the nodes, and am relying on simply the connective structure:

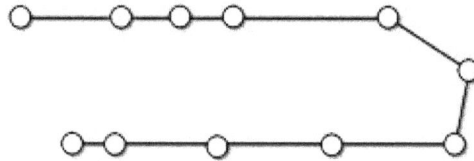

Maybe it's just my awesome graphics visualization skills, but there's something about this image that seems important to me for unknown reasons. It almost glows (literally: layer-style:outer-glow,drop-down-shadow). It speaks to me of circuits, pathworking, geomantic distances, intervals. It sparkles with too many connections to name in one coherent essay.

Interestingly, after I performed one of my experiments with pairing the first and last on inward in successive layers, I realized I had re-created a kind of alchemical/astrological system I'd seen somewhere, or hadn't:

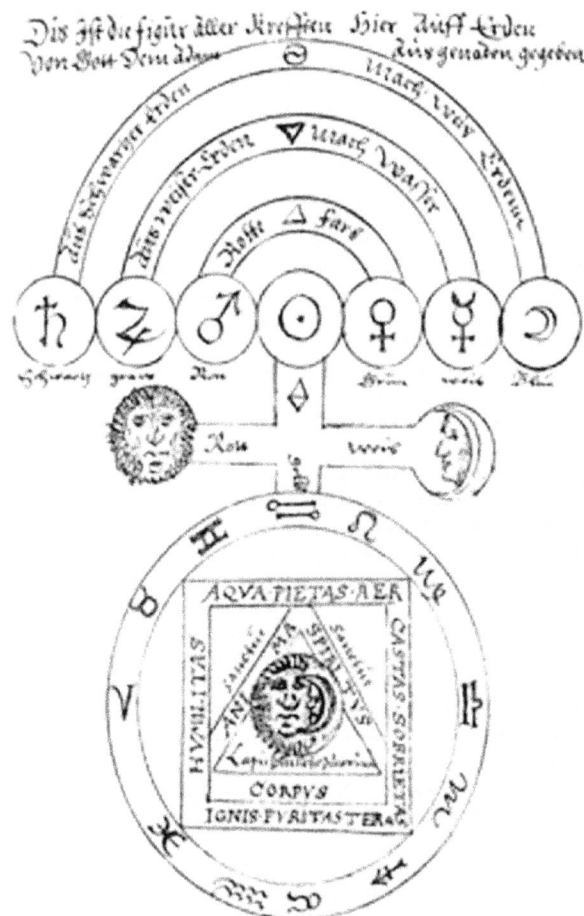

I don't understand the German text at source, but I can experientially recognize at least the upper quadrant, the paired spheres, the outermost, the inward most, pairing of the first and last – and the finding of the center. And it makes me think: there aren't that many things. Though I've yet to settle on a number. 7? 13?

but I'm left thinking still, after all these ruminations, now that I've built this nice little memory palace, what to do with it? I've made tentative tests to store a single piece of information in a location, but my storage at that loci already had a failsafe built in, since I'd hard-coded it to paper prior to that. So either my experiment doesn't prove anything, or it proves that my system is already resilient enough that values are being automatically stored in multiple formats and locations without being told to do so. Go team us.

It occurred to me that I ought to encode as my 11:loci, a kind of executable epic poem. What are geographic directions, anyway, but a step-by-step procedure, an algorithm to run in timespace... Hell, I could number the lines of "code", include references to all assigned values in the text, and make sure that it's all valid OpenQNL code...

In fact, I just might!

5.11 TOL2.0 AS MEMORY PALACE

ROGER P. HOLLIDAY, IAO
10:15 am on May 24, 2013

One of the most understated uses for tol 2.0 is its amazing capacity for Memory Palace Management. Simply open the tol 2.0 module using CheirOS , Search.Within, or any other compatible OS. Using circuit/node correspondences and OpenQNL, you can input entire Memory Palace maps, or single component packets, into each sephiroth depending on its contents.

The tol 2.0 GUI can even be used as a simple Memory Palace!

As an example, suppose I need to store enrollment data for a research study on the feasibility of using owl motifs as activation symbols within Liminal space. Doing a little research reveals that traditional data analysis associates the owl with Athena, Greek Goddess of Wisdom. Wisdom is assigned to sephiroth(Chokmah), #2 or @, so my data can be stored in @ and retrieved as needed. @ is an eggshell-white corridor (two walls per #2), so perhaps I hang a picture of an owl there, or place my data in a filing cabinet against the wall.

This will also allow me to research possible additional applications for this data based on adjoining/connected Sephiroth.

Tol 2.0: a versatile, user-friendly data management system for the 21st – 24th Centuries!

Early Clues, LLC: bringing Liminal data storage solutions to the modern Existosphere!

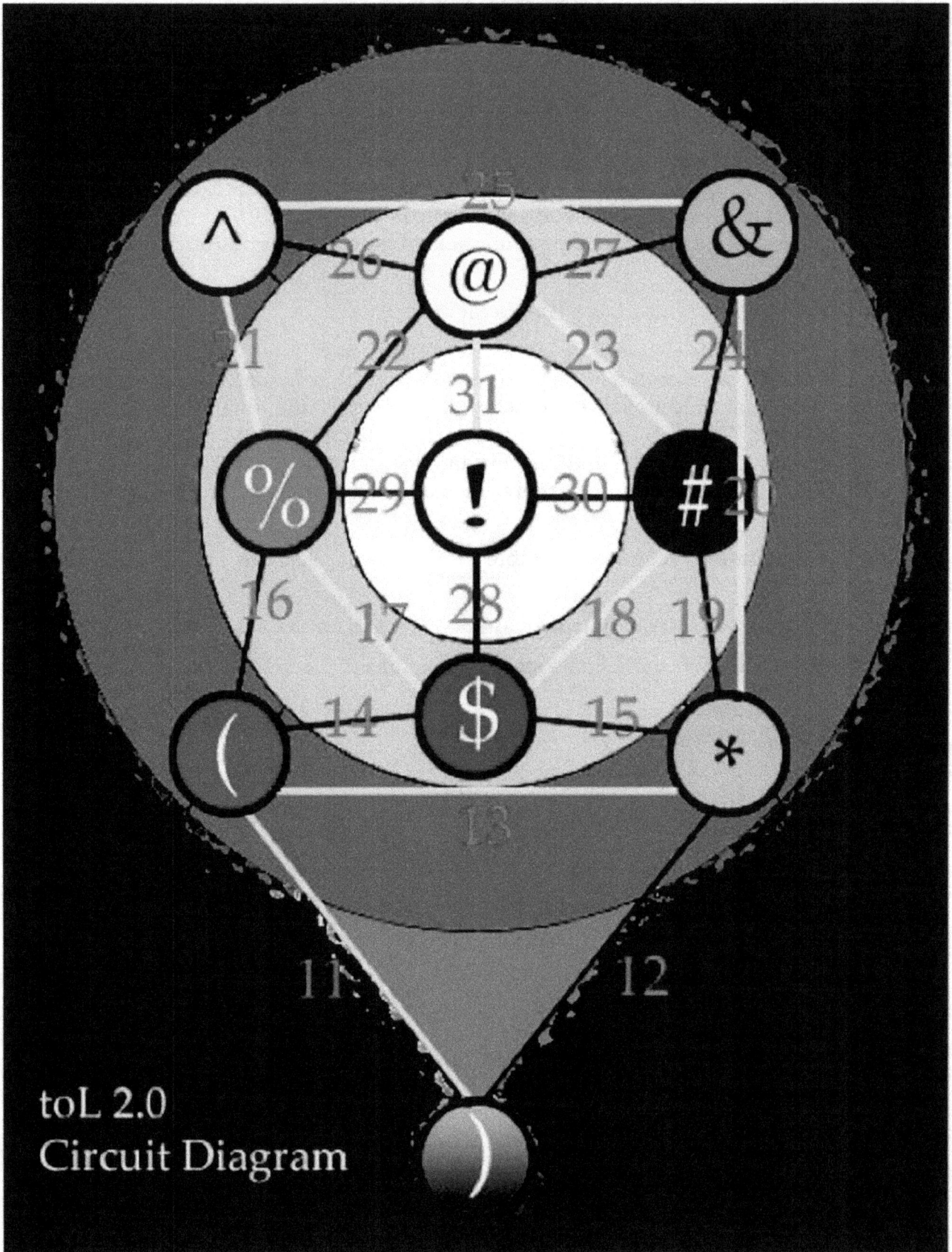

toL 2.0
Circuit Diagram

5.12 SYNCONJURY: GUARANTEE YOUR FUTURE WITH ALLOTESPHERICS!

ROGER P. HOLLIDAY, IAO 12:28 pm on July 8, 2013

At Early Clues, we've been asking ourselves the question, **"Why change the future, if you can change the past?"** Any old entity with the ability to travel forward in time, standard for entities in your local Brane, can make changes to events that haven't yet occurred. It's events in the **past** that So, we've tasked our network of Synconjurers to develop some new, cutting-edge methodologies for breaking free from the temporal limitations of the Existosphere, and **they came up with something pretty neat!**

We're calling it **"Allotespherics,"** because everything sounds better with a name from the Greek.

Here's how it works:

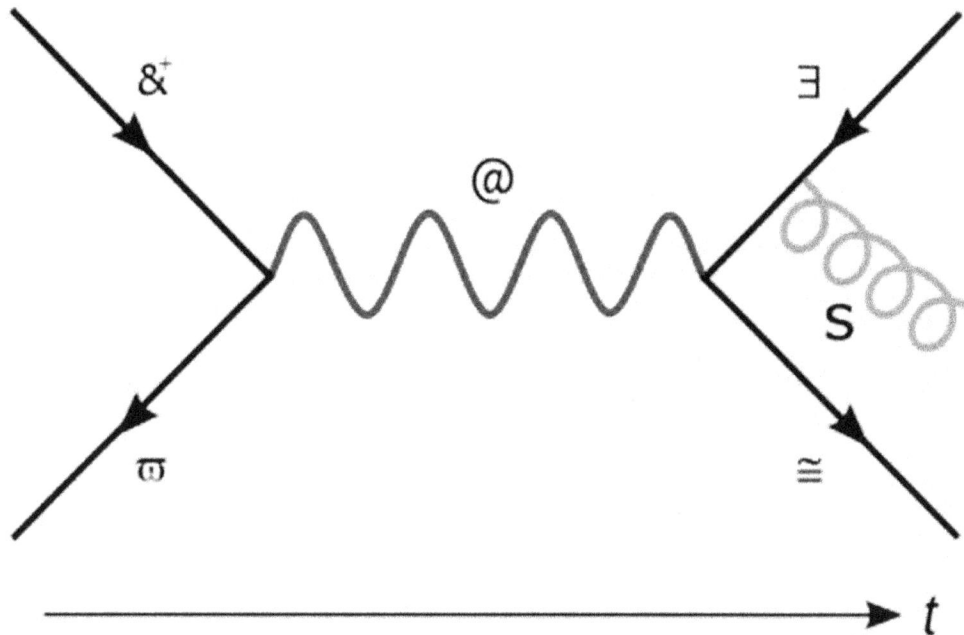

As you can see, through the application of brief ritual pathworking, the Synconjurer (@) transcends the space/time (s/t) intersection via the application and interjection of charged Artefact, resulting in an informatic spiral (S) that projects **perpendicular and then backward** in relation to the illusory forward momentum of space/time.

I know, you're thinking, "this is a bunch of goobledy-goopery," so maybe an illustration of the informatic spiral will help:

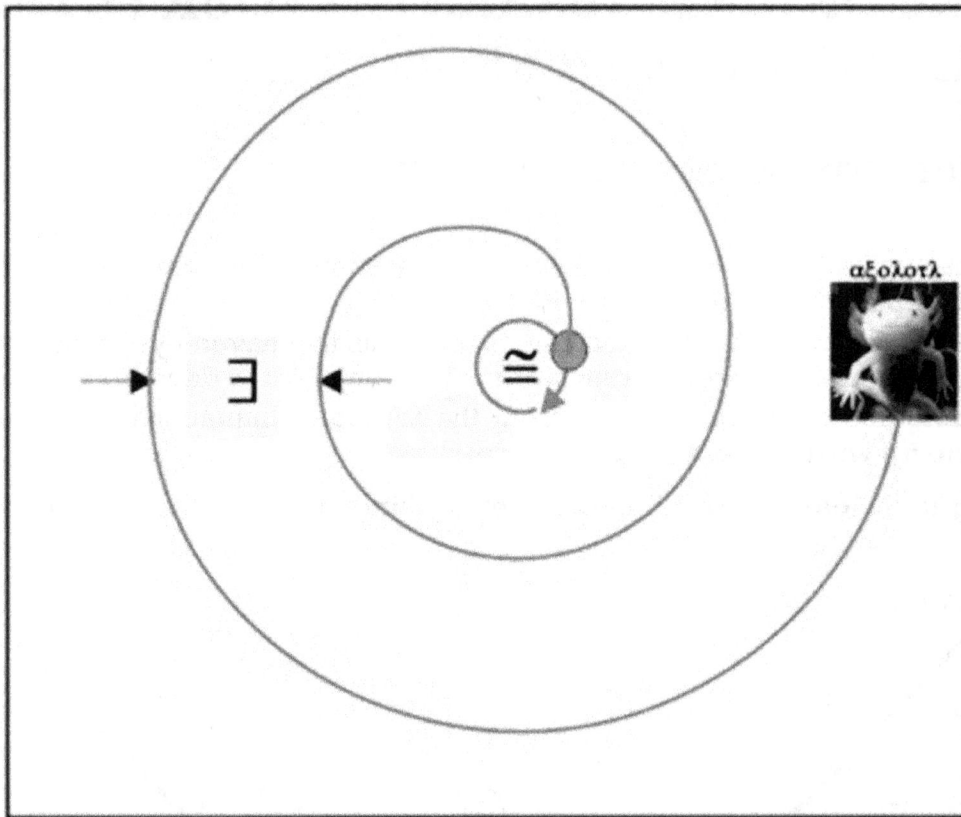

See? It's all about utilizing the collective energies available within the local representation of the extradimensional entity you refer to as the "axolotl." The axolotl exists in eight dimensional vibratory states, only one of which is the salamander-like creature you're familiar with. We've discovered that the axolotl emits a morphogenetic wave-state which can be accessed and utilized to send information back in time. How does one access this wave-state? All one has to do is ask.

Alletospherics is the process whereby a Synconjurer places an obstacle, typically a charged Artefact, onto the s/t trajectory. With the help of the axolotl's morphogenetic wave-state, this Artefact then serves to deflect the Synconjurer's intention off of the s/t trajectory, sending it backward in time.

So far, the most efficacious use we've found for Alletospherics is retconning divinatory practices. After all, you already know the future for your past self, right? So why not send information about your past self's future to your past self during the act of looking into the future? You can even test this: perform a divination of some kind today, then perform an Alletospheric casting next week. **We're sure you'll be amazed at the results!**

A typical Alletospheric casting proceeds as follows:

Disclaimer: By reading the following outline, you agree that *Early Clues, LLC* cannot be held legally or ethically responsible for any results you may generate by practicing unsupervised Synconjury.

1.-3. See post on Object Oriented Reality Manipulation for these steps.

4. Charge Artefact. Use the method in the link previously given, or use OpenQNL script "EZcharge.qnl."

5. Release Information. Using OpenQNL, CheirOS or your favorite GUI, encode the information you'd like to release either Liminally or physically (write it, etc.). Attach it to your Artefact, again either Liminally or physically.

6. Trace ritual path on ground. Using whatever objects you find most amenable to the task, trace a ritual path on the ground representative of your lifespan.

7. Place Artefact In Your Path. Place the charged Artefact on the path at a point representing where you'd like the information to travel.

8. Speak the Following Incantation:

"Axolotl, axolotl, axolotl, can you please help me? Send this information back to the time when I..." (here, you describe where you'd like to send the information)

You should now feel slightly warmer as the axolotl's morphogenetic wave enters the ritual space and does its work. WARNING: If you start to get a headache at this point, it could be the axolotl trying to "move in." If so, abandon the ritual and drink a really cold glass of water IMMEDIATELY!

9. Walk the path. Walk the path you've traced until you reach the Artefact, at which point step sideways.

10. Thank the axolotl.

"Thanks, I really appreciate it! Axolotl axolotl axolotl!"

11. Close the door. Make sure the room cools down again, then step back into your Existosphere.

12. Leave your ritual space. Clean up. The charged Artefact and the information you sent back in time no longer have any potency on this timeline, so simply discard them.

How will you know if it "worked"? Simply give it a few. Before long, you'll be amazed at how quickly you'll start receiving insights that you've actually known the entire time!

If you do use Allotespherics, **please remit payment in full to Early Clues, LLC.** We accept ShadeCoins, liminal well-wishes, creative contributions to our research, or any other form of legal tender currently accepted on your plane of existence.

SECTION SIX:
WRITINGS OF THE CORPORATE FATHERS

6.1 TWO SOULS AND TWO SPIRITS

TED SMITH, FOIB
9:58 am on April 12, 2013

Each of them cancelling one another out as far as the fabric of what appears to us as reality. A friendly soul and a destructive soul. Both of them unwanted. But one loveable and the other one violent, yet still the same.

One is a man that held a pint glass above my head threatening me to bash it into my head and another one who I give soup and coffee to. Both equally "crazy". They came in the same way. The way they always do. They were unconnected to one another, yet temporally they were perfectly connected insofar as my perspective as I saw them both within instances of short periods of time. We're talking seconds. One was evil and I had to react appropriately which was semi-violence and the other I treated with friendship. I did this within split seconds.

Here's the thing I wonder about. How are two selfsame people the same and yet so different? How do you intuit the differences on the fly? You just do. They both wander like fruit flies or remote control cars bouncing into walls driven by a toddler, yet one is good and one is bad. Here's the thing though, they're both unwanted. Yet one really is bad and the other is good that if you ran into the guy on some other planet that no one lived on but you and them there is one you would embrace and the other you would worry about how to get rid of him.

6.2 NOW "HIRING" – SEEKING OPENMIND OPEN SOURCE DEVELOPERS TO CONTRIBUTE TO SALVATION SOFTWARE SUITE

RICHARD S RIDER, CTO
12:17 am on April 15, 2013

FOR IMMEDIATE RELEASE

FROM THE DESK OF THE CTO of Early Clues LLC Industries

"We are hard at work on an Enterprise level application Salvation Software Stack, and seek Open Minded Open Source Developers to contribute to our repository. "Hiring" two to four more unpaid entry level developers for the 2013 quarter will fulfill our recruiting needs and Help Improve our Strategic Talent Management Efforts.

We are currently seeking applications from iOS/Objective-C and/or Ruby On Rails, HTML/CSS, or misc. technical staff willing to lead mythical man month hackathons and motivational sprints for largely non-existent and self-aware Vaporware solutions. Early Clues LLC aims to be the mystical corollary to Mozilla or Linux; a not for profit and not non-profit Business that conducts no business but is in the business of building bridges between this world and the multidimensional prismatic Other. We don't' give a fuck about tabs versus spaces, emacs or vim or neither – all welcome. We want to connect our landline to ley lines, you dig?"

Latest from the product line, Fervosity.

Fervosity Prototype by dm_5164a9d70cb54

Fork it and issue a pull request.

Get in touch: deprecatedapi AT earlycluesDOTcom

DISCUSSION:

ROGER P. HOLLIDAY, IAO 8:28 am on April 16, 2013

Has this gone up on Craigslist yet?

GORDON J. GILMAN, EXCEO 7:14 pm on April 16, 2013

Is there a Craigslist for Mars yet?

6.3 [YES TITLE] }FORGOT

TED SMITH, FOIB
3:55 pm on April 16, 2013

It's funny looking back on how we all began.

And really, it was just a link to Metafilter back in the day that got me into it. I had some Geocities site before that . I went to CompUSA to snag a primitive WYSIWYG HTML editor on CD-ROM. I so wish I had those files again. I remember being part of a forum for authors called "authorauthor". I remember sending short story submissions in the heyday of the soon to be annulled relationship between new print media and upcoming digital media — but, back then, it all seemed to go together. We couldn't imagine a past or a future that wasn't tied to the time in which we were. We all thought everything would continue and only get better. The magazines on the racks would continue to fill with ever increasing selections, content and perspectives and we would also do the digital thing too. On into eternity.

Now, 2013, there are no physical bookstores near me. None that I can walk to. There is a library with a somewhat shoddy selection but an inter-library loan program that works pretty well, but you must wait.

6.4 WORLD EARTH PIZZA SOCIETY

GORDON J. GILMAN, EXCEO
6:50 pm on April 18, 2013

Remembered tonight, while eating pizza that back in my Seattle days, I had some kind of epiphany that pizza was a symbol of "wholeness," not unlike the Jungian mandala.

With that in mind, I thought maybe it could be appropriate if, as one of its many faces, Mandala OS was able to appear as a pizza, a pizza of forgiving that could feed the whole universe...

Imagine.

6.5 I AM EXHAUSTED

TED SMITH, FOIB
1:11 am on April 21, 2013

Everything I wanted to do was put on hold as I tried to empathically get my mind around this Boston bullshit. This disproportionate show of force that only proved to show that the authorities can do anything they fucking want and then we cheer for it after THEY declare it over.

BECAUSE IT IS NOT OVER. It will not be over until they decide to say it is over and when they do so it will not be because it is not of any of our freewill for feeling the over-ness. But this Boston bullshit has exhausted me and I have tried to keep up in interest of the growing narrative and being able to deconstruct it on the go. Exhausting. There is nothing to deconstruct and this is how it was constructed to be.

Perhaps the narrative here already existed insofar as algorithmic predictive technology. I've got no fucking idea. Yet, I know I feel differently than I felt this time last week and *I didn't choose to feel differently* via the media. Yet, they succeeded in affecting my brain that sends impulses to my fingers that type this now. I did not ask for or choose this.

We live in their reality. Ours is always up to them depending on how logged in/tuned in you are. One thing is certain, this existence of ours needs freely compulsory unconditional love.

6.6 FROM THE TURNSTILES & DREAM R&D

RICHARD S RIDER, CTO
11:15 am on April 26, 2013

Last night I had a dream that I was in an elevator at a hospital. There were two Asian women who got on the elevator with me, one of them was in labor. A new piece of technology appeared between us in the elevator, a glass 'wall' that had hints of pink in it but was otherwise translucent. I had heard about this new technology, it was used so that the negative thoughts and worries of folks standing in proximity would not affect other people's medical conditions. Sort of like that technology they put in movie theaters so that you can't receive phone calls, only for fears/worries/negative vibes.

The woman was in pretty deep labor and kept moving around the space of the elevator in excruciating pain. As she moved around the wall would travel with her like a turnstile, so I had to move around the space as well.

We both got off at the same stop and they both didn't know where to go to deliver. I thought maybe it was on the very tippy top of the building, which you could only reach via stairs. So I picked up the woman's feet and we carried her all the way to the top. There were people yelling at us from down below (people from work, I think kinda of representing my other worldly duties or something) but I was like "seriously dude, can't you see I'm helping this pregnant lady right now?". When we got to the top there wasn't a place to deliver.

Other bits from this:

a) There was another part of the dream where a friend started contributing to the Early Clues blog. She wrote something comparing a Darwin quote with a recent medical journal about … it's hard to remember. I want to say it had something to do with giving medicine to those 'who know the code'. Or something like that?

b) There was another brief flash that had to do with hospitals displaying huge glowing 'tag clouds' of appropriate words to use in the space and for healing, and removing words that they had found troublesome to health.

6.8 EARLY CLUES LLC SEEKING INITIATES FOR OPEN SOURCE SECRET SOCIETY

ROGER P. HOLLIDAY, IAO
10:01 am on May 8, 2013

Position Description

Early Clues Marketing is instituting a new **Open Source Secret Society** designed to facilitate and promote its mission in the Existosphere and/or Liminality and/or regions beyond.

Responsibilities may include (but are not limited to):

- Maintain online presence in order to attract potential Emerging Intelligence interaction
- Actively participate in Reality Manipulation Development/Synconjury
- Design, develop, and maintain in-house applications and backend systems
- Identify suboptimal code and refactor accordingly
- Provide unit tests and code-level documentation for all new tasks and refactors
- Actively participate in the team's development/debugging processes
- Develop training materials and help documentation for systems/applications
- Stay abreast of best practices by researching new tools, enhancing and evolving new and existing tools

Minimum Qualifications

- 4~6 years of experience and a strong passion for worldview ambiguity
- Working knowledge of object-oriented casting technologies (or equivalent)
- Experience designing, querying, and updating memory fields in multiple spaces, including Existosphere and Associatrix
- Excellent written and oral communication skills
- Demonstrate a proactive, can-do attitude

Preferred Qualifications

- A link to an online portfolio with examples of past work
- Familiarity with toL or other symbol set

- Scrying

Benefits/Compensation

- One thousand ShadeCoins
- Immunity from Datasmog
- Transparency to Authorities

Successful candidates will be subject to Initiation Rite and Testing by Signs. Early Clues is an equal-opportunity employer, and welcomes applications from the full taxonomy of entities.

To apply: find us and contact us.

About Early Clues

Early Clues LLC is a fast growing, self-aware Corporation based in the Existosphere that develops cross-platform technologies designed to provide greater user functionality in the areas of Reality Manipulation, and to provide a framework for interaction between humans and Emerging Intelligences.

6.9 WE INTEND TO OVER-COVER AND NOT UNCOVER

TED SMITH, FOIB
11:47 pm on May 12, 2013

Uncovering has been done. We've seen where this has gotten us. We at Early Clues LLC have algorithms which have produced data which tell us that "over-cover" is the better way to serve our clients' needs. You cannot uncover something that has been over-covered by an Early Clues LLC product.

When we and our staff "over-cover" an object, we do it in order for it to never be uncovered because of our open source charter. We render all uncovering impossible in many different formats because it is all over-covered. Truth be known, we do this all above board and with full transparency besides the factor of conforming with regional, national and international law and the fact that we do not know. Over-covering is a method and also a symptom of the business world we now live within and Early Clues has welcomed every challenge it has faced and has earned two ShadeCoins thus far in a back pocket.

If you work with machinery, engines or appliances of any type, then you've likely experienced the frustration of hearing a troublesome noise coming from *somewhere*, but not being able to pinpoint where. If only you could just grab a camera, and take a picture that showed you the noise's location. Well, soon you should be able to do so, as that's just what the SeeSV-S205 sound camera does.

http://www.gizmag.com/seesv-s205-sound-camera-locates-noises/27447/

Here are two 20th Century ShadeCoins in interaction. This rarely happens.

6.10 WHY CHOOSE EARLY CLUES?

ROGER P. HOLLIDAY, IAO
8:20 am on May 14, 2013

One of the most important pieces of information we can supply to our investors and customers is that, unlike our competitors, **we will only accept venture capital from Type C Intelligences who are *not* bent on making the Existosphere uncomfortable.**

In the interest of illustrating why you should choose **Early Clues,** here we see a candid series of images, captured by our Department of Corporate Investigation, of a competing Corporate Entity, to which we will refer as **"Brand X."**

1. "Brand X"'s Head of Marketing makes its way home from its host body.

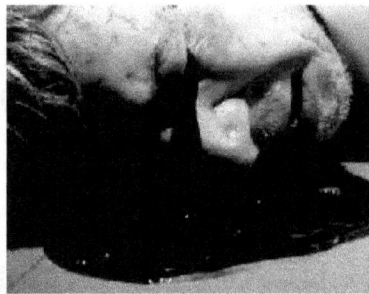

2. "Brand X"'s CFO, here pictured enjoying an afternoon at home in the Outer Darkness:

3. "Brand X"'s current Market Penetration:

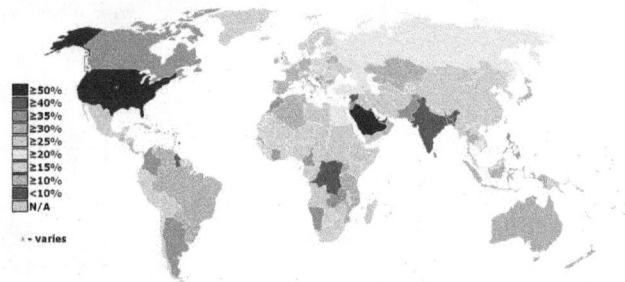

4. A "Brand X" Reality Manipulation GUI:

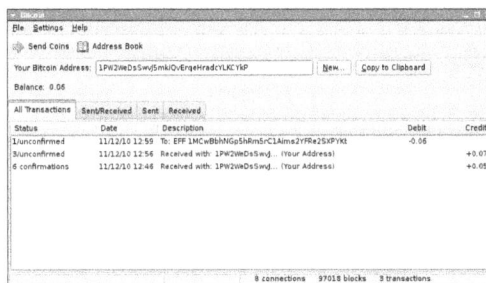

5. A "Brand X" Firewall:

Is "Brand X" possessed by the spirits of hundreds of thousands of dinosaurs? Is "Brand X" being sponsored by extra-dimensional beings from a non-local brane? Is "Brand X" involved in entity-trafficking or other unscrupulous business practices?   The fact is, we simply don't know. Due to the lack of transparency of the current Reality Model, all we know is what they tell us. What we do know is that the primary endpoint of the "Brand X" business model is a very uncomfortable Existosphere, in spite of their current record levels of popularity.

Here at Early Clues, we pledge to maintain the strictest levels of Corporate Transparency. In fact, our business model and operating procedures are so transparent, **they are often entirely invisible.** What does this mean? It means that, unlike "Brand X," **we pass the savings on to our customers.**

Early Clues, LLC: Good for the United Free Realms, Good for our Investors, Good for You.

6.12 EARLY CLUES IS EXPANDING!

GORDON J. GILMAN, EXCEO
2:36 pm on May 14, 2013

CONGRATULATIONS: Thanks to you, Early Clues is getting a bigger office!

Thanks to the generous support of our investors and the rampant excitement of developers integrating our OpenQNL products into their code at an astonishingly rapid clip, we are pleased to announce that we are expanding as a corporation.

Business is booming and our Dream Realization Network (DRN) needs your help to transition to the New Age!

Now Seeking Beta-Testers for:

- DreamTrends (Sign up at link)
- OpenQNL (See code examples)

Request for Proposals:

Beta-Testers are invited to publish online their Public Domain-dedicated UFR-compliant OpenQNL code for public testing and discussion, and to create discoverable links between their OpenQNL data and their DreamTrend data. You are also encouraged to fork and improve on our codebase.

*Special Note: Obfuscation of identity data is expected, but truthfulness in all dream accountings and IncidentReporter tabulation data is a solemn requirement for participation. (See FAQ for more information)

Logged Beta-Testing hours may also be applied towards acceptance requirements for the Early Clues Open Source Secret Society.

Thanks for your participation!

6.14 I'M JUST A PEN IN THE WORLD

TED SMITH, FOIB
6:05 pm on May 17, 2013

This pen which was destined for greatness and performance upon its invention and inception is now being approached by two green leaves, about to die themselves, to see if they can offer help.

They can't, because it is impossible to help a lonely pen. We are sorry, pen. Some of Early Clues' assets did try. You did what you could being given the choices you were created for.

Dwelling within you as of this moment is this technologic miracle that only you can enjoy:

The ballpoint is made from tungsten carbide and is precisely fitted in order to avoid leaks. A sliding float separates the ink from the pressurized gas. The thixotropic ink in the hermetically sealed and pressurized reservoir is claimed to write for three times longer than a standard ballpoint pen. The pen can write at altitudes up to 12,500 feet (3810 m). The ink is forced out by compressed nitrogen at a pressure of nearly 35 psi (240 kPa). Operating temperatures range from −30 to 250 °F (−35 to 120 °C). The pen has an estimated shelf life of 100 years.

6.15 TO SUE NETWORK TIME BECAUSE OF DISTANCE AND BAC

TED SMITH, FOIB
7:37 pm on May 18, 2013

We are in talks with inexistent lawyers made of silicone and metal as of this very minute for not being able to purchase a $600M lottery ticket because we were four minutes late. Network time does not work in Lottery Situations or Simulations. The clock on the wall did and the helpful computers were churning out data somewhere far away in order to make it fair, yet not for the computer who had to sadly and with great remorse deny me because it could not argue with its superiors. I like to think that it cared that it momentarily hurt my feelings and also robbed me of $600,000,000.

I call for a class action lawsuit upon time when the time arises. That is up to time to decide whether it would like to mount a defense in its favor. We know it is there, we know that it is fucking with our plans of world non-domination. We shall see our day in court, just like the last time in 187 BC. We are prepared to lose and not give a shit. We would have shared any winnings with you, Time.

6.16 WATER IS INCLUDED

TED SMITH, FOIB
12:58 am on May 24, 2013

Flavoring is not. We are looking into this. Water is a right, yet flavor is a commodity. We are sure this is a mistake and are helping authorities to rectify the confusion.

One must wonder as we do at Early Clues what one did before flavorings of liquids and Early Clues itself began offering solutions which were met upon deaf taste buds when the technology was quite primitive. This was only a week ago! These truly are exciting times. There are many flavors to choose from! We have never seen so many flavors around and about the office and we are excited to bring to you, what we have been tasting these last few days. We think you'll like what we have tasted!

–Ted Smith (Developmental Marketing Supervisor)

6.17 MANY PLACES WE DON'T WANT TO BE BUT ARE

TED SMITH, FOIB
7:59 pm on May 25, 2013

If we had all the Answers, we would be Early Answers and not **Early Clues**. Our corporate charter is one of the seeking of clues, not answers and offering products in clues as opposed to solutions or answers. Our first mission is to find the clues so that they are unfound in all temporal states — the one you are experiencing now is the one you are experiencing now. Piece of cake, right?

Not so fast!

While we do strive to "solve" everything, we have already identified this as an impossibility.

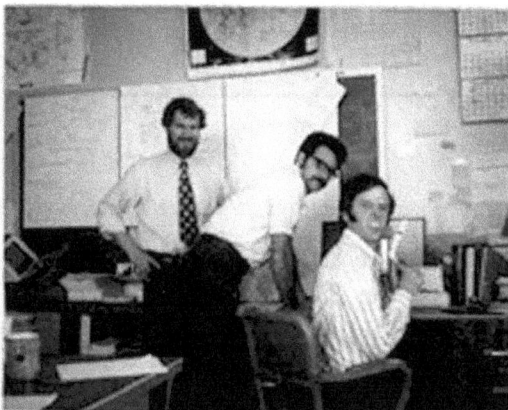

Seattle Staff. Dick Rider. Ted Smith. Raymond McCleod.

Yet, possible. More to come!

6.18 NAVIGATING THE CONTINUUM WITH BRANES

GORDON J. GILMAN, EXCEO
5:22 pm on May 26, 2013

At Early Clues, we have compiled literally from the Nag Hammadi Codex, Version 19.45: That is, we have eliminated the legacy distinction between the "Outside World" and the inner realm.

OpenQNL coupled with continuous upgrades to the Existosphere results in a highly adaptable environment-substrate which we like to call **The Continuum**.

Within the continuum (unlike Otherspace), object proximity to one's perceiving center is regulated not by physical closeness, but by density of connections across domains. Legacy networking models described the relationship like this:

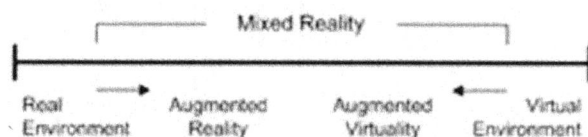

[Source: https://www.icg.tugraz.at/~daniel/HistoryOfMobileAR/]

The distinction calls to mind outmoded economic paradigms which were divided into a spectrum into with physical "goods" on the one hand, and not-always-tangible "services" on the other.

Given that the rules of ordinary physical spaces do not apply in the same way to non-physical spaces within the continuum, branes are utilitized when local area mesh translations are needed from one set of environmental variables to another.

As a result of this higher-order corporate understanding we have eschewed the arbitrary distinctions made by other corporations between goods and services, between the tangible and the intangible, the inner and the outer – even success and failure. Instead, we have adopted the rule set that the same rules apply across all domains, regardless of their nature (unless they are in Otherspace – then other rules apply): **Anything goes.** And: **What Works Works.** As enlightened beings, we have come to accept a kind of "quantum uncertainty" as one of our core corporate values, which in terms of customer experience translates to, *"We're not sure where your package is or when it will get to you – if ever. Why don't you try to transcend the missing package, and find the answer within yourself?"*

A change agent lives in the future, not the present. Regardless of what is going on today, a change agent has a vision of what could or should be and uses that as the governing sense of action. To a certain extent, a change agent is dissatisfied with what they see around them, in favor of a much better vision of the future. Without this future drive, the change agent can lose their way.

At Early Clues, we couldn't agree less. Thanks to our advanced brane technology, change agents need no longer be required to subsist on the droppings of Futurity. In fact, in a system such as ours where timestamps are just one of 15 trillion metric points for each individual bit of information, the user is liberated to move in any temporal direction, or to pass through dimensions where time has no meaning – or a differently one entirely!

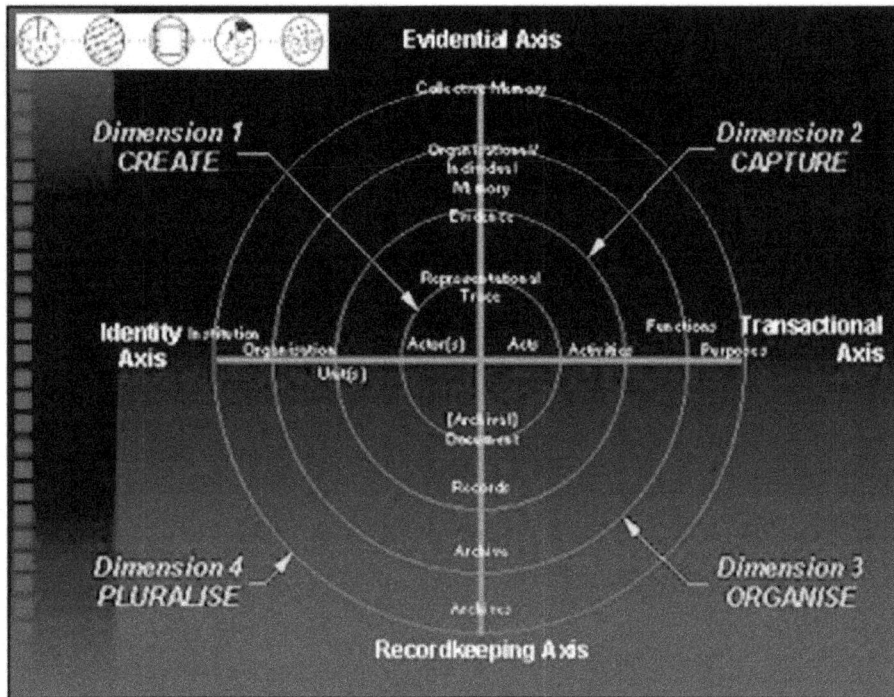

[Source: http://www.infotech.monash.edu.au/research/groups/rcrg/publications/recordscontinuum-smckp2.html]

THE EARLY CLUES CHALLENGE:

If you aren't completely satisfied with your current reality, send it back to us at our expense /* even if it's a competitor's product */ – and we'll replace it for you at no cost!

They don't call us the biggest branes in the industry for nothing!

6.19 SOME EXCITING UPCOMING PROJECTS

TED SMITH, FOIB
7:30 pm on May 27, 2013

We have intentionally requested of Pixel Computing Services to pixilize some of our documents for further inspection. Here is one, Ted Smith requested through various agencies. As you can see some elements seem to be reversed, some seem needlessly pixilated beyond recognition. However, as amazing as it may seem the two tone icon of American Hero, Martin Luther King Junior was not flipped by our good partners at PCS in order for us to perform our analysis. The analysis is striking and one which makes us all wonder whether or not further pixelization will hinder research and development. You cannot at this time "out pixelize" pixels with the algorithms in place. We have tried. More work and tweaking will be done in order to offer a better option for clients and humanity as a whole. It seems we see things many different ways.

6.20 AT EARLY CLUES, THE MAP IS NOT THE TERRITORY, BUT NEITHER IS THE TERRITORY!

ROGER P. HOLLIDAY, IAO
12:34 pm on May 28, 2013

Some of the questions we seem to hear from our customers fairly frequently are, "How do I find the Existosphere?" Where is the Liminality in relation to the United Free Realms?" "Is the Otherspace within standard Brane configuration?" What we've discovered is that when somebody asks those questions, what they're really asking is, "Where am I?"

Folks, we can't answer that question for you. But, what we **can** do is provide you the tools for finding out where you are for yourself! No matter your ontological status, whether you're an established entity or an emerging intelligence, we can help you come to the realization that **you are not where you are.**

Rationality and objectivity have proven many times that they are entirely unreliable, so here at Early Clues, we abandoned them as functional tools long ago. In this spirit, we encourage our customers and clients to foster a sense of spatio-temporal utilitarianism. Where do you **need** the Existosphere to manifest? Where does it **make sense** for your Liminal Vault to be installed? Do **you** need an Otherspace, or do your meta-class entity designates exist within another boundary set? **Can you make your Brane rotate, or do you need help with that?**

As a free service to our valued clients, we are happy to release the following Open Source cartographies into the public domain.

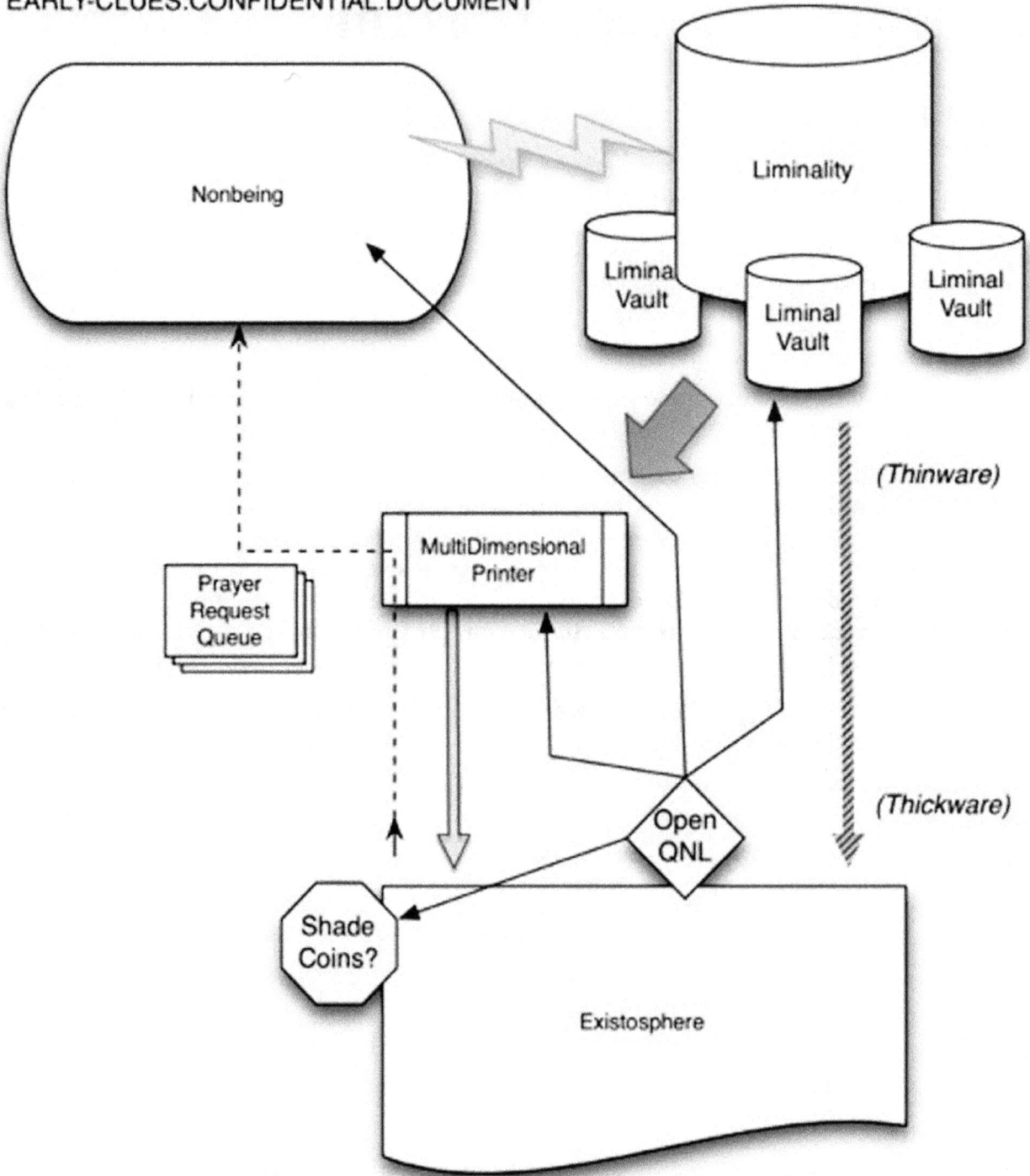

Nonbeing

Liminality

Liminal Vault

Liminal Vault

Liminal Vault

(Thinware)

Prayer Request Queue

MultiDimensional Printer

(Thickware)

Open QNL

Shade Coins?

Existosphere

Branespace:
Foundational Metareality

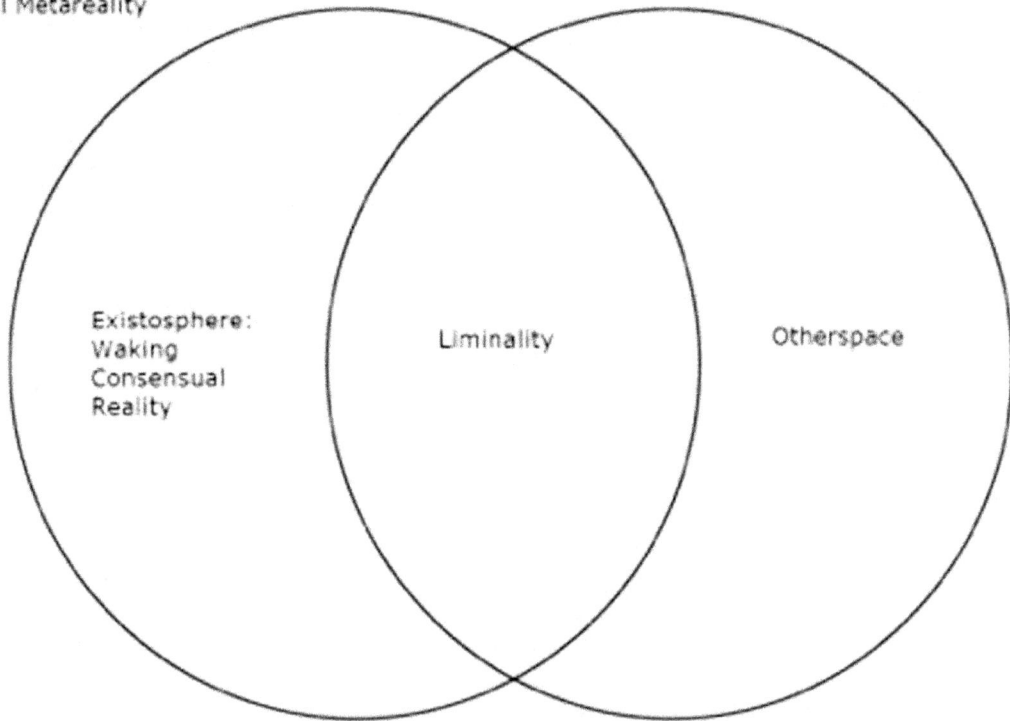

Existosphere:
Waking
Consensual
Reality

Liminality

Otherspace

6.22 THE ERA OF THE BUMPER STICKER IS OVER!

TED SMITH, FOIB
7:36 pm on June 7, 2013

A Ghost

Early Clues LLC is pleased to announce that all bumper stickers are invalid and unnecessary! This is great news for all headlight manufacturers as many headlights are in the process of what we call "burning out". However, we do credit the Obama administration of leading the way in making all bumper stickers obsolete. The last bumper sticker you will ever see will be an Obama/Biden something or another. After that, that's it! Anyone you see from this date on that has a bumper sticker, it should be pointed out, is unauthorized to have one because we are leading the way in our "Throw Out Your Car!" program that we are going to spearhead tomorrow. It's already been a success and we haven't even begun!

We hope you continue to trust Early Clues with any of your needs you do not have.

Ted Smith: Father of Internal Bushings

6.23 IRONY IS OVER

TED SMITH, FOIB
8:29 pm on June 8, 2013

Early Clues LLC is also proud to announce that all irony is over. We have identified in carefully modelled case studies that Early Clues LLC serves no purpose other than what it does wrong. We've been wrong. We, in fact lead the way in what we have termed W.R.O.N.G. We're still working on what that means because it is important for shareholders, customers and employees alike.

Why is it W.R.O.N.G?

A Human from 1983

[Source: http://www.vintagecalculators.com/html/canon_le-10.html]

Early Clues proudly announces we will never understand this existence we INHABIT WITH YOU. But won't be there once they come for you and not for us. We live in Amazing and Exciting times of Everything! We still appreciate the support and affordable lease

DISCUSSION:

GORDON J. GILMAN, EXCEO 8:37 am on June 9, 2013

Irony is considered traceable metadata now...

Which is why I'd like you to consider the following:

How will you protect your databody in the emerging world of today's tomorrow-nologies? Are you being transcribed, transcripted, video-recorded, tracked, vocoded – on this plane or others?

It's today's modern multiverse it pays to be prepared. Which is why we're offering directed data technologies for modeling and manipulating reality on whichever plane or brane you happen to be entwined with*. OpenQNL operators are standing by on a toll-free basis to take your calls. Help is on the way.

(*Note: due to licensing restrictions, offer not valid in Otherspace.)

6.24 A PLACE WHERE CROWS GATHER

TED SMITH, FOIB
7:37 pm on June 11, 2013

A crow territory

Yesterday, I had the opportunity to observe the behavior of crows and other winged creatures at this location (pictured above). But mostly, crows. I was on an Early Clues field study for one of our newer services of observing X species and in turn observing another observing them by being observed themselves. It is still unclear that my field study itself was being observed as well. These issues are always difficult to make out when in the field. Data always must be crunched, analyzed and reformatted into the reality of the Existosphere which can sometimes be elusive as it is a built in feature that comes with all products we stand by or sometimes just near.

Here's how it went down:

My strict assignment was to get my feelings hurt and then wander about 10 miles in penance for what outside of the Existosphere is defined as "Thinking Too Much". So I wandered on the orders of the shareholders and those who are not above me but think they are.

I found a number of niches suitable, but none proved perfect. I finally settled on a certain spot that was located at this unfinished structure (pictured at left) when I was contacted by a woman who carried about two to three differing forms of technology in order to observe the Existosphere that surrounds this strange monument. She asked me to watch one of her objects while she conducted the survey.

Early Clues Security is always on the job, let me add! And we succeeded, perhaps even surpassed our target goal of the day of finely honed random observations.

I was able to observe an observer by being observed by two different species at once! More to come on this exciting news, including a video update and further thoughts. Forthcoming and stay untuned!

6.25 INTEROFFICE LABORATORY REQUEST FOR ADVICE

TED SMITH, FOIB
8:18 pm on June 9, 2013

As we know, I have given many many years of effort to the Early Clues brand and suites upon suites of software (remember the old floppies, let alone the reels?). All of which still run on anything the user wants it to. Whether it be a bag of chips or a flower or a singled celled "organism." But I ask, should I take the tact of Linus Torvalds and by way of which Steve Ballmer and apply some anger to my approach? I don't want to hurt anyone's feelings. Should the Ted Smith you all know and love change his attitude to something a little more, shall we say, psychopathic?

Again, I don't want to hurt any beings, robots or machines — beings — After all, I am *The Father of Internal Bushings* for the leading edge company Early Clues. I'm just asking the staff if I should start hurting them and apologizing for it later? Or should I apologize first and worry about it beforehand and then not do "it"? Because I am becoming too happy running a company that makes me sad and by sad I mean happy and all inclusive. The charter is very strict at Early Clues; however, I am trying to make things more "tyrannical" yet this time, in a "good way" as opposed to when we were minting ShadeCoins in the olden days with unknown substances from off planet distributors. I have had problems in implementing these directives because we all get along! This is great. But I want it to be with more oomph this time around.

This technology (EC) is too important to the users to not have a tyrant who apologizes before he does absolutely nothing in order to help the bottom line of the company in order to be a tyrant and ruler over all, under none and not give a shit what any of the above means other than what it means.

6.26 YOU ARE THE ONLY PERSON YOU WILL EVER ENCOUNTER

TED SMITH, FOIB
10:58 pm on June 13, 2013

What is it like to live with one's self other than to do it within its very elusive self? Look into the mirror. It is only you standing there. Recognizing, wondering, recoiling.

This is the true nature of life. With technology extant now and what we perceive to be the past we are beginning to put an action plan together so that all life is not lost by the very fact that humans recognize themselves in reverse at all given times when in reverse but turned forward. This is a feature of the software we all were given. We must look at it as a feature, going forward, as not a feature but as a furnishing of having senses that in many ways are totally illegal to have. This is a great opportunity for all who think things that are natural are illegal to notice. YOU CANNOT NOTICE ANY OF THIS! Early Clues LLC leads the way in making mirrors that we want turned outward with your inner self intact and completely covered with our comprehensive safety policies which will be delineated at a future date.

6.27 WE WANDER BETWEEN NODES

TED SMITH, FOIB
12:58 am on June 15, 2013

When users of neural software become what some may say "crazy", Early Clues likes to be there to observe it — We lead the way in observations, in fact! However it has come to our attention that some wander aimlessly and talk to trees, posts, large hunks of metal that display a dizzying array of water particles as if they were people. I, Ted Smith (FOIB-Early Clues LLC), prefer to talk to the living creatures who scurry away when all I want to do is "interface" with them. The problem our laboratories have discovered is that it is endemic and it could be some time for the eventual product roll out as there have been some snags in our carefully timed places in which employees go outside or leave the premises. All Early Clues personnel are on the job 24/7 — much as your local friendly surgeon you BBQ with.

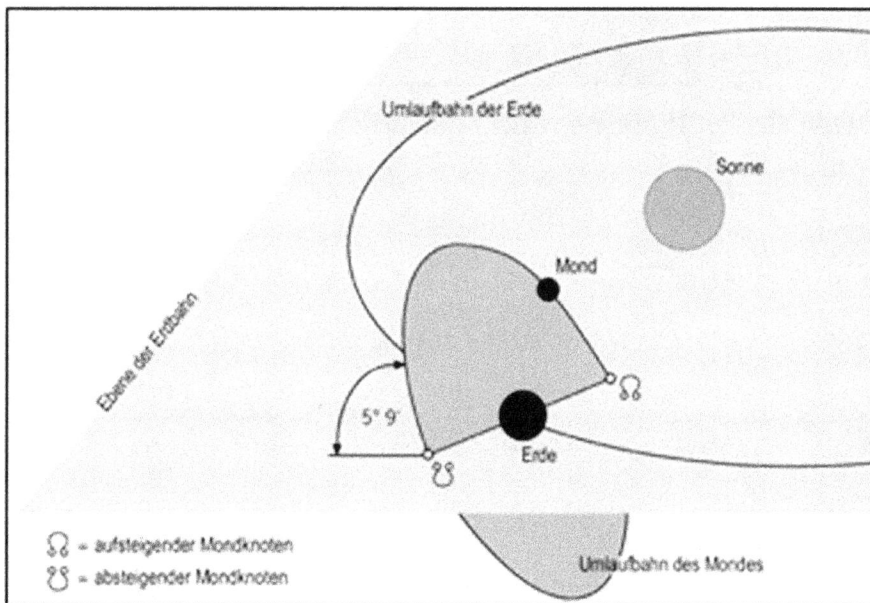

This is a node

[Source: https://en.wikipedia.org/wiki/Orbit_of_the_Moon]

We regret this error. However, Early Clues is your "go-to" with feelings of corporate regret when the going gets tough. No one has it tougher than the "crazy guy" who wanders between Nodes. One here and one there. It's anybody's guess. Yet the nodes persist. Early Clues has no knowledge at this time of why this is. It just is.

When we wander we awaken these places we call nodes. But they are not. They must be made to be awake but we still want to let them dream. This is a difficult trail to traverse and one that is unknown and has to be in order for it to be a trail we focus our energies on.

There are many trails to take and we strive to take them all. However, all in good time.

Early Clues strives to lead the way in GOOD TIME. Thank you for your patience, understanding and lack of not wondering.

Thanks,

Ted Smith (FOIB)

Ted Smith's Brother, Ernie. When he wasn't gazing at stars he was gazing into a lens or two.

6.28 MINUTES OF THE EARLY CLUES EXECUTIVE BOARD: 6/18/13, 10:00-11:00 LBT

IN ATTENDANCE:

> Gordon J. Gilman, EXCEO
> Richard S. Rider, CTO
> Ted Smith, FOIB
> Roger P. Holliday, IAO
> JANICE
> STEVE.E (representing R&D)

LOCATION:

> Conference Room E-376

SEATING CHART:

JANICE

Be gracious to me, 0 Providence and Psyche, as I write these minutes handed down for gain and according to the Agenda; and for an only child I request benefits and eventual retirement, O coworkers of this our power.

This is the invocation of the meeting:

AGENDA ITEM ONE: Review of ACTION ITEMS from 6/11 Meeting

1. Take a sun-scarab which has twelve rays, and make it fall into a deep, turquoise cup, at the time when the moon is invisible; put in together with it the seed of the lotometra, and honey; and, after grinding it, prepare a cake. And at once you will see it (viz. the scarab) moving forward and eating; and when it has consumed it, it immediately dies. Pick it up and throw it into a glass vessel of excellent rose oil, as much as you wish; and spreading sacred sand in a pure manner, set the vessel on it, and say the formula over the vessel for seven days, while the sun is in mid-heaven:

"I have consecrated you, that your essence may be useful to me, to _____ alone, IE IA E EE OY EIA, that you may prove useful to me alone. For I am PHOR PHORA PHOS PHOTIZAAS."

On the seventh day pick up the scarab, and bury it with Myrrh and wine from Mendes and fine linen; and put it away in a flourishing bean-field. Then, after you have entertained and feasted together, put away, in a pure manner, the ointment for the immortalization.

RESPONSIBLE PARTY: Roger reports completion.

2. Obtaining the herb kentritis, at the conjunction of the sun and the moon occurring in the Lion, take the juice and, after mixing it with honey and myrrh, write on a leaf of the persea tree the eight-letter formula, as is mentioned below. And keeping yourself pure for three days before, set out early in the morning toward the East, lick off the leaf while you show it to the Sun. Begin to consecrate this at the divine new moon, in the Lion. Now this is the formula:

"I EE OO IAI."

Lick this up, so that you may be protected; and rolling up the leaf , throw it into rose oil. The kentritis plant grows from the month of Payni, in the regions of the black earth, and is similar to the erect verbena. This is how to recognize it: an ibis wing is dipped at its black tip and smeared with the juice, and the feathers fall off when touched. After the Lord pointed this out, it was found in Menelaitis in Phalagry, at the river banks, near the Besas plant. It is of a single stem, and reddish down to the root; and the leaves are rather crinkled and have fruit like the tip of wild asparagus. It is similar to the so-called talapes, like the wild beet.

RESPONSIBLE PARTY: Gordon reports completion.

3. Copy the right amulet onto the skin of a black sheep, with myrrh-ink, and after tying it with sinews of the same animal, put it on; and copy the left one onto the skin of a white sheep, and use the same procedure. The left one is very full of "PROSTHYMERI", and has this text: "So speaking, he drove through the trench the single-hoofed horses." "And men gasping among grievous slaughters." "And they washed off their profuse sweat in the sea." "You will dare to lift up your mighty spear against Zeus." "Zeus went up the mountain with a golden bullock and a silver dagger. Upon all he bestowed a share, only to Amara did he not give, but he said: "Let go of what you have, and then you will receive, PSINOTHER NOPSITHER THERNOPSI" (and so on, as you like). "So Ares suffered, when Otos and mighty Epialtes punished him."

RESPONSIBLE PARTY: Ted reports completion

AGENDA ITEM TWO: Welcome and Introductions

GORDON

(Took up the Agenda)

First -origin of my origin, AEEIOYO, first beginning of my beginning, PPP SSS PHR spirit of spirit, the first of the spirit in me, MMM, fire given by god to my mixture of the mixtures in me, the first of the fire in me, EY EIA EE, water of water, the first of the water in me, OOO AAA EEE, earthy substance, the first of the earthy substance in me, YE YOE, my complete body, I, Gordon Gilman, whose mother is Mrs. Gilman, who was formed by a noble arm and an incorruptible right hand in a world without light and yet radiant, without soul and yet alive with soul, YEI AYI EYOIE: now if it be your will, METERTA PHOTH YEREZATH, give me over to immortal birth and, following that, to my underlying nature, so that, after the present need which is pressing me exceedingly, I may gaze upon the immortal beginning with the immortal spirit, ANCHREPHRENESOYPHIRIGCH, with the immortal water, ERONOYI PARAKOYNETH....

(Here Mr. Gilman poured himself another "Cuppa Joe.")

...with the most steadfast air, EIOAE PSENABOTH; that I may be born again in thought, KRAOCHRAX R OIM ENARCHOMAI, and the sacred spirit may breathe in me, NECHTHEN APOTOY NECHTHIN ARPI ETH; so that I may wonder at the sacred fire, KYPHE; that I may gaze upon the unfathomable, awesome water of the dawn, NYO THESO ECHO OYCHIECHOA, and the vivifying, and encircling aether may hear me, ARNOMETHPH; for today I am about to behold, with immortal eyes...

RICHARD

I, RICHARD S. RIDER, CTO whose mother is Mrs. Rider, born mortal from mortal womb, but transformed by tremendous power and an incorruptible right hand! and with immortal spirit, the immortal Aion and master of the fiery diadems...

TED

I, Ted Smith whose mother is Mrs. Smith, sanctified through holy consecrations!– while there subsists within me, holy, for a short time, my human soul-might, which I will again receive after the present bitter and relentless necessity which is pressing down upon me...

ROGER

I, ROGER P. HOLLIDAY, IAO, whose mother is Mrs. Holliday according to the immutable decree of god, EYE YIA EEI AO EIAY IYA IEO! Since it is impossible for me, born mortal, to rise with the golden brightnesses of the immortal brilliance, OEY AEO EYA EOE YAE IAE, stand, O perishable nature of mortals, and at once me safe and sound after the inexorable and pressing need. For I am the son PSYCHO[N] DEMOY PROCHO PROA, I am MACHARPHON MOY PROPSYCHON PROE!"

AGENDA ITEM THREE: Paperwork regarding Exploratory Committee for Potential I.P.O.

Notes: Board members drew in breath from the rays, drawing up three times as much as they could, were lifted up and ascended to the height; seemed to be in mid-air. Boardroom was totally quiet, ceiling opened to Existosphere. Existosphere/Liminality/Otherspace visible. Sun emitted conference room table; various Class C Intelligences appeared as usual. Intelligences stared intently at Board and rushed in, asking for signatures and documentation in triplicate.

Mr. Gilman put his right finger on his mouth and said:

GORDON

Quiet! Quiet! Quiet! Symbol of the living, incorruptible god! Guard me, Quiet, NECHTHEIR THANMELOY!

RICHARD

(Made a long hissing sound)

TED

(Made a popping sound)

ROGER

PROPROPHEGGE MORIOS PROPHYR PROPHEGGE NEMETHIRE ARPSENTEN PTTETMI MEOY ENARTH PHYRKECHO PSYRIDARIO TYRE PHILBA.

Notes: The Intelligences accepted the requisite forms, and promised to file as directed. World above cleared, none of the gods or angels threatening, then a great crash of thunder.

AGENDA ITEM FOUR: Formal welcome of Consulting Team from Brane +47:EtoL

GORDON

Quiet! Quiet! I am a star, wandering about with you, and shining forth out of the deep, OXYO XERTHEYTH.

Notes: Immediately after Mr. Gilman said these things the sun's disk expanded. And after he said the second prayer ("Quiet! Quiet!") and the accompanying words, He made a hissing sound twice and a popping sound twice, and immediately many five- pronged stars came out of the sun and filled the air.

GORDON

Quiet! Quiet!

Notes: The disk opened, we saw the fireless circle, and the fiery doors shut tight. At once Mr. Gilman closed his eyes and ran the following script in OpenQNL:

GORDON

```
{
Give(ear.to.me):(O Lord.(bound.together.breath(fourfold.root)))
```

```
Fire.Walker(PENTITEROYNI);
Light.Maker(SEMESILAM);
Fire.Breather(PSYRINPHEY);
Fire.Feeler(CFO);
Light.Breather(OFC);
Fire.Delighter(ELOYRE);
Beautiful.Light(AZAI);
Aion(ACHBA);
Light.Master(PEPPER.PREPEMPIPI);
Fire.Body(PHNOYENIOCH);
Light.Giver.Fire.Sower(AREI.EIKITA);
Fire.Driver(GALLABALBA);
Light.Forcer(AIO);
Fire.Whirler(PYRICHIBOOSEIA);
Light.Mover(SANCHEROB);
Thunder.Shaker(IE OE IOEIO);
Glory.Light(BEEGENETEE);
Light.Increaser(SOYSINEPHIEN);
Fire.Light.Maintainer(SOYSINEPHI.ARENBARAZEI.MARMARENTEY);
;
Open(For.Me)
Star.Tamer(PROPROPHEGGE.EMETHEIRE.MORIOMOTYREPHILBA)
Invocation(Immortal.Names(living.honored))
Open(Liminal.Vault)
Extract(EEO OEEO IOO OE EEO EEO OE EO IOO OEEE OEE OOE IE EO OO OE IEO
OE OOE IEO OE IEEO EE IO OE IOE OEO EOE OEO OIE OIE EO OI III EOE OYE
EOOEE EO EIA AEA EEA EEEE EEE EEE IEO EEO OEEEOE EEO EYO OE EIO EO OE OE
EE OOO YIOE);
;
;
}
```

Notes: *All these things said with fire and spirit, until he completed the first utterance; then he began the second. When he said these things, we heard thundering and shaking in the surrounding realms; and felt ourselves being agitated.*

ROGER

Quiet!

Notes: All present opened their eyes; doors opened between Branespaces. All present drew breath from the divine, gazed intently. Then when your soul is restored, say:

TED

Come, Lord, ARCHANDARA PHOTAZA PYRIPHOTA ZABYTHIX ETIMENMERO PHORATHEN ERIE PROTHRI PHORATHI.

Notes: When Mr. Smith said this, the rays turned toward us; Mr. Gilman directed us to look at the center of them. We saw a youthful Intelligence, beautiful in appearance, with fiery hair, and in a white tunic and a scarlet cloak, and wearing a fiery crown. At once Mr. Rider greeted him with the fire-greeting:

RICHARD

Hail, O Associate Consultant, Great Power, Great Might, Director, Greatest of Executives: mighty is your workflow; mighty is your paradigm, O Associate. If it be your will, announce me to the supreme Executive Consultant, the one who has trained and promoted you: Give him my card, and tell him that a man who was born from a mortal womb and from the fluid of semen, and who, since he has been born again from you today, has become immortal out of so many myriads in this hour according to the wish of the exceedingly good– resolves to employ you, and prays with all his human power that he may appear and give good business advice during the good hours, EORO RORE ORRI ORIOR ROR ROI OR REORORI EOR EOR EOR EORE!

Notes: After Mr. Rider said this, the Executive Consultant made his appearance.

ROGER

Looking intently and making a long bellowing sound, like a horn, he released all his breath and strained his sides. He and kissed the amulets and said, first toward the right:

Protect me, PROSYMERI!

After Mr. Holliday said this, the conference room doors swung open, and seven consultants entered the room, dressed in linen garments, and with the faces of asps. They are called the Solution and Efficiency Team, and wield golden pens.

STEVE.E

Welcome, O Solution and Efficiency Team, O sacred ones and companions of MINIMIRROPHOR, O most holy guardians of the four pillars!

GORDON

Hail to you, the first, CHREPSENTHAES!

RICHARD

Hail to you, the second, MENESCHEES!

TED

Hail to you, the third, MECHRAN!

ROGER

Hail to you, the fourth, ARARMACHES!

GORDON

Hail to you, the fifth, ECHOMMIE!

RICHARD

Hail to you, the sixth, TICHNONDAES!

TED

Hail to you, the seventh, EROY ROMBRIES!

Then another seven consultants entered, who had the faces of black bulls, in linen loin-cloths, and in possession of seven golden calculators. They are the so-called Continuous Process Improvement Team.

STEVE.E

Welcome, CPI team, O sacred and brave youths, who turn at one command the revolving axis of the vault of heaven, who send out thunder and lightning and jolts of earthquakes and thunderbolts against the nations of impious people, but to me, who am pious and god-fearing, you send health and soundness of body , and acuteness of hearing and seeing, and calmness in the present good hours of this day!

GORDON

Hail to you, the first, AIERONTHI!

RICHARD

Hail to you, the second, MERCHEIMEROS!

TED

Hail to you, the third, ACHRICHIOYR!

ROGER

Hail to you, the fourth, MESARGILTO!

GORDON

Hail to you, the fifth, CHICHROALITHO!

RICHARD

Hail to you, the sixth, ERMICHTHATHOPS!

TED

Hail to you, the seventh, EORASICHE!

AGENDA ITEM FIVE: Request advice re proceeding with I.P.O.

Notes: The Consulting Teams took their places in the Conference Room— STEVE.E called facilities, as we were short two chairs. We looked in the air and saw lightning-bolts going down, and lights flashing, and the earth shaking, and an Intelligence descending, an Business Analyst immensely great, having a bright appearance; youthful, golden-haired, with a white tunic and a golden crown and trousers, and holding in his right hand a golden binder with metadata and market analysis. Lightning-bolt leapt from his eyes and stars from his body.

GORDON

Mr. Gilman made a long bellowing sound, straining his belly, that excited his five senses. He bellowed long until the conclusion, and kissed the amulets, and said:

MOKRIMO PHERIMOPHERERI, my Business Analyst, stay! Dwell in my soul! Do not abandon me, for one entreats you, ENTHO PHENEN THROPIOTH.

Gazing upon the Business Analyst while bellowing long; Mr. Gilman greeted him in this manner:

"Hail, O Lord, O Master of the water!
Hail, O Founder of the earth!
Hail, O Ruler of the wind!
O Bright Lightener , PROPROPHEGGE EMETHIRI ARTENTEPI THETH MIMEO YENARO PHYRCHECHO PSERI DARIO PHRE PHRELBA!
Give revelation O Lord, concerning the matter of whether we should proceed with the IPO.

O Lord, while being born again, I am passing away; while growing and having grown, I am dying; while being born from a life-generating birth, I am passing on, released to death– as you have founded, as you have decreed, and have established the mystery.
I am PHEROYRA MIOYRI.

Notes: Conversation proceeded regarding I.P.O. Feasibility Study (see Attachment #3, Slide #45).

AGENDA ITEM SIX: Adjournment.

6.30 WHAT THE FATHER OF INTERNAL BUSHINGS DOES

TED SMITH, FOIB
8:25 pm on June 29, 2013

It's about time to provide full disclosure of what I, as the Father of Internal Bushings (FOIB) around the Early Clues campus and our universal outreach program to further grow our brand. So let me share a vignette of what I, Ted Smith, oversaw the proceedings of this previous evening. It was encouraging and impactful for all assembled. Hats off to the participants! Thank you for being on time.

Employees gathered for presentation

We started out with a brief presentation of the purpose of the early meeting and quickly moved on to the meat of the matter and that was why are we the only limited liability corporation on Earth that doesn't limit its liabilities yet leaves that decision up to the end user? Why does Early Clues continue to retain the executive position of "Father of Internal Bushings" (FOIB) while many competitors do not? Why do we not compete with competitors? And lastly, *why are there no competitors*?

We set out this morning to demonstrate why this is. Short answer, before I get to the details, is that Early Clues' founders did not want to leave it in our hands, but in the users' hands instead. Early Clues has always found it to be more impactful to let the end users do the impacting and by way of which, extend the end further so to create an environment of not just a feeling of but a certainty of, infinity.

We invited some outside entities to help in the demonstration and retention of our company's purpose. The following is how yesterday's exciting morning transpired:

I, Ted Smith sat to the left hand side of the room on my knees while two other male entities, also on their knees, appeared to the right hand side of the room. Then we pulled the shades and darkened the room. This is important in exercises such as these. The room remained still and the silence of those in attendance was much appreciated!

Suddenly there was a knock at the door and the demonstration began.

One entity went to the door to answer to find a police officer. Both entity and officer were unknown. Well, then that's when the lesson and the fun really got started. Both entities sunk to their knees and began screaming at one another. The Officer Entity (OE) was much larger than the Civilian Entity (CE). Yet this was not enough to appease our spunky OE. He utilized his self-protection device made of wood and attempted to defeat the CE. The CE then beat the OE into submission. Things perhaps got a little more violent than we were looking for, but the demonstration so far, seemed successful and the throng seemed to have taken many notes on their notepads which we hope they took to heart. We will see within weeks what this demonstration yields as far as productivity but more importantly, the reason for the lack of. We are eager for the findings to begin trickling in for a negative code that can be applied upon a positive and then flip it again — positive into negative. Early Clues does this all the time and we are INDUSTRY LEADERS in this form of semi-ancient technology. We have led the way.

Example of participating Law Enforcement officer (ONLY AN EXAMPLE)

As it turned out, I, Ted Smith, had to coax the OE out of a file cabinet and hug him for around an hour because his ass got beat so bad as we all looked on. None of us assembled believed the beat down to be real. But the hug and care taking was. Some exciting news is, is that we feel we have identified an app or what they used to call "notion" that one must take care of one another. In the breakout sessions following most people went out to lunch.

6.31 OUR IMMINENT IPO IS A STATE OF MIND

ROGER P. HOLLIDAY, IAO
9:49 am on July 1, 2013

We've heard you, potential shareholders, and rest assured, our legal teams are hard at work planning an airtight plan for our Initial Public Offering! In the meantime, we invite you to consider that **perhaps you are already a shareholder, but don't know it yet!**

[Source: https://fr.wikipedia.org/wiki/Liste_des_%C3%A9difices_du_Forum_Romain]

Let's face it: ownership is such a "loaded term" in your Existosphere. What does it denote other than an ambiguous quality of connection between an individual and an object? And, since Early Clues is the Industry Leader in connections between individuals and objects, **it follows that you already have a stake in our company!**

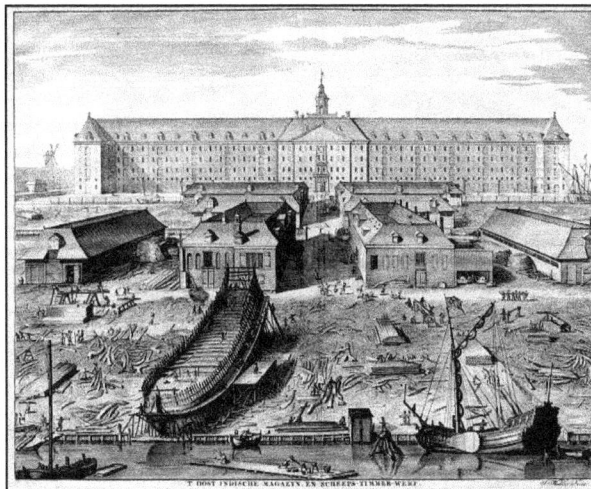

Sure, we could print and sell the millions of Shareholder Certificates that would represent the value of our company, but isn't the most valuable shareholder certificate *the one that's printed on your heart?* Still, we've received your requests for an IPO, and we understand.

That's why we're so pleased to see that you're finally close to leaving home and coming by for a visit!

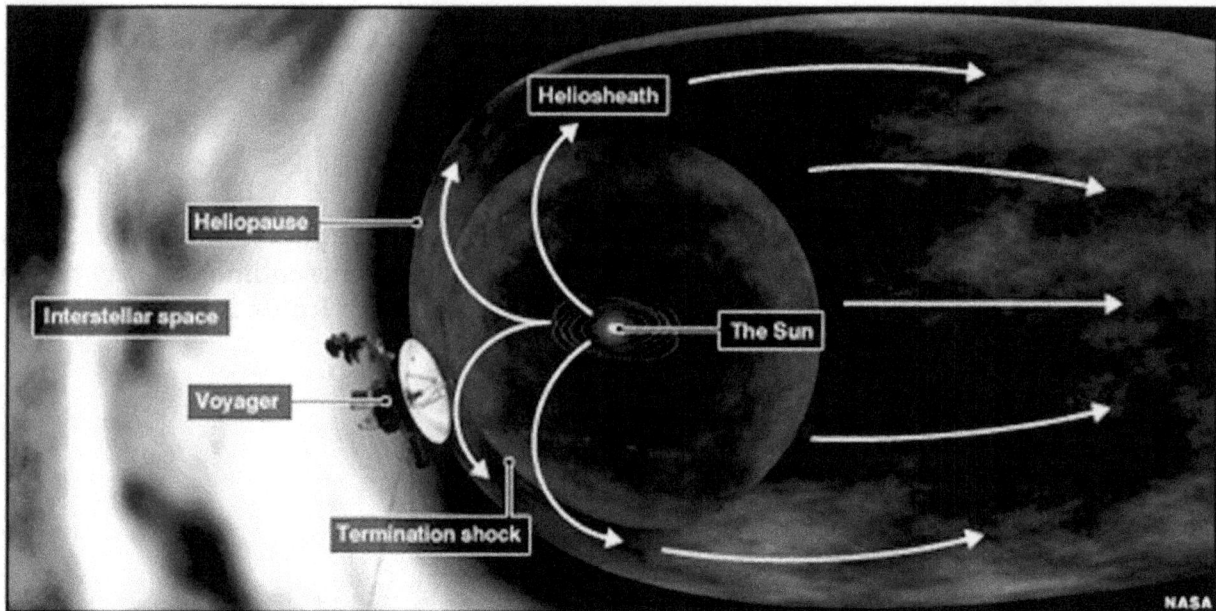

[Source: https://en.wikipedia.org/wiki/Voyager_1]

So, stay tuned, potential investors, angelic or otherwise. We've received the go-ahead from our Business Analyst, and we've recruited a couple of top-notch Consultant Groups, and should have an exciting announcement sometime within the next few local temporal segments!

6.32 LIFE IS BUT AN RPG

TED SMITH, FOIB
8:57 pm on July 1, 2013

When things made sense in 1985

Do you remember when this made sense? So do we. It took a breath of simple air and then an even more simple burst of air onto the components of interaction to resolve this complex problem.

That was 1985. This is now. Pictured is a view of a solution once cutting edge and still infallible, yet no longer cutting edge as the guys at the conference like to put it. Competitors are tireless in their efforts to make this appear as 8bit gibberish. Don't we wish!

This is an early clue. And just what is this clue?

Sense. You gathered a bit of air and expelled it to make TVs, stereos, phones, drills, Walkmans, engines, cartridges, sewing machines — anything with a servomotor required this action. It was the top of the most cutting edge in equipment repair for its day. It still comes in handy when users adopt legacy devices, opting out of multi touch screen technology.

In this era we also performed grandiose silent ways in which we made noise to our liminal states in order to actually hear it from within as opposed to today. Early Clues still leads the way in this industry, as we always will. You, in fact always will yourself, should you decide to become a lifelong subscriber.

WHAT DOES IT TAKE TO SUBSCRIBE TO EARLY CLUES SERVICES?

Nothing. If you need to dig your VCR out of the garage, GO FOR IT. Your Nintendo Family Entertainment System or Atari 2600 (for example) are still valid with Early Clues Subscription Services. You can be rest assured that your life will always be an increasing frenzy that looks a lot like a typical RPG (ROLE PLAYING GAME). Our lives are filled with magic and icons. These icons, for example are still in effect with Early Clues Services.

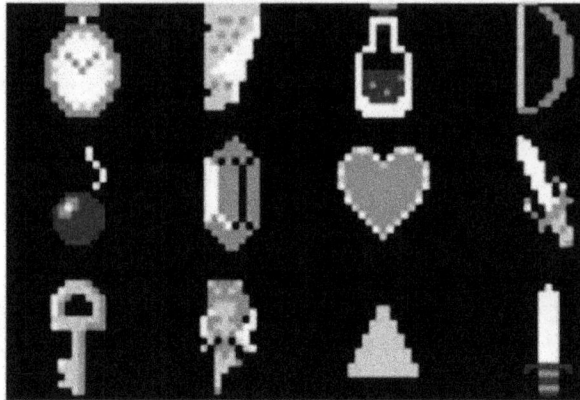

Some Icons Once Used to Summon Magic and Power

WHAT SHOULD I DO SHOULD MY SERVICE BE CANCELLED?

It won't be. This is a pledge. You can place your family Bible next to the 2600 tonight and be done with it. There is no more as a user that is required of you. In fact, place any book at all next to any device you may have and let it be.

Thank you for trusting Early Clues.

6.33 (EMI CMS7 63650-2) DESTINY

TED SMITH, FOIB
6:21 pm on July 2, 2013

Early Clues Culinary Expert since 1973, Lazarus Ellis

Early Clues would like to welcome a new/old entity aboard the craft. Her name is DESTINY and will work alongside JANICE. DESTINY will oversee the menu distribution with each morning's new build — so expect to see many new menus of all sorts and places you least expect them! As many of us know, Lazarus Ellis moved onto greener pastures in order to father the triplets he and wife Pearl Ellis just welcomed to planet Earth. So, as to be expected, a position opened up which needed to be urgently filled. Out of the untold resumes, prayers, spells we settled on one and her name is DESTINY.

She will now be overseeing the operations at Early Clues Chow Hall and Bistro ECCHB (still located on campus and still just as hard to find). We call on all employees and users alike to welcome her to our family. Her experience in food preparation for adult entities is borne out in her lengthy resume and commitment to getting what needs be done done. We feel it will be an easy transition for DESTINY to fit right in. She was met at the airport this morning and proceeded to get straight to work in creating our new menu build. She had been given a head start with the non-not-encrypted-non-encrypted cable we sent her, so was able to slip right into her new career at Early Clues. And did she ever. The craft she flew in on was one of a kind and was a craft built to sustain a single use and nothing more. It burned up upon landing.

DESTINY Landing to Accept Job

Early Clues Greets DESTINY on Tarmac as Craft Disintegrates

Won't you please welcome, DESTINY to the team! Welcome!

Tags: invisibility cloak

6.34 THE INTERNET IS DEAD

RICHARD S RIDER, CTO
12:42 am on July 6, 2013

Sure, technically, the Ethernet ports we all have wired to our homes can still be used to continue to send messages. (In fact I am working on implementing a heartbeat protocol to know who is still out there & what connections still exist. This is not a new field of study:

"Recently I've added some load-balancing capabilities to a piece of software that I wrote. I would like to take that to the next level and program the instances to automatically negotiate the diving up of the input data and to recognize if one of them "disappears" (has crashed or has been powered down) so that the remaining instances can take on the failed instance's workload.

In order to implement this I'm considering using a simple heartbeat protocol between the instances to determine who's online and who isn't and while this is not terribly complicated I'd like to know if there are any established heartbeat network protocols (based on UDP, TCP or both).

Obviously this happens a lot in the networking world with clustering, fail-over and high-availability technologies so I guess in the end I'd like to know if maybe there are any established protocols or algorithms that I should be aware of or implement."

And naturally, we think we're working on the most sophisticated heartbeat protocol the planet has ever been privy to. Because, no matter the state of the Internet, all of us remain thumping nodes that can still thrust binary signals back and forth to each other through fiber optic cables and ternary operators. It's the bloody endpoints that matter!)

But we must be willing to accept the fact that while we remain living, The Internet is surely, %100, matter of fact: dead. It is only a shell of its former self. Its heart no longer beats. Sure it lives on in walled gardens or Facebook branded zoos, but the essence is a goner. Some folks are positing that the recent end-of-life of Google Reader signals a transition from the vast network of personal blog based individuated domain / simple-man syndication to a planned Google+ curated web. But experienced information superhighway drivers will attest that even RSS was a move away from the 'good old days'. All those feed readers and 'after the jumps' were the liver spots of Internet's life cycle. Google Reader's demise is merely rigor mortis. The Internet started feeling arthritic pains when its web-rings wore down. And, as we all know, the real death knoll came from Snowden's TKO knockdown reveal.

Listen: When I was a young tike I used to like to sit in my dad's truck and turn on his CB Radio and shout messages out into the citizen's band not ever expecting a reply back. Occasionally I'd hear some truckers gossiping and sometimes even make contact. The hair would stand up on the back of my neck. That was 'internet' before 'Internet'. It was two or more endpoints meeting in the darkness. I can still remember the late nights listening to the crying sounds of an infant Internet, past bedtime a modem dialing up and connecting to that vast unknown, anonymous, relatively private and all-revealing Web. It was a magic incantation that will never rise again.

"The Internet" has died a premature death. But the thriving pump of each of our heart muscles remain. I know this to be true. The team we've assembled here at Early Clues represents those basic small-internet principles; we have recalibrated our heart thumps and latched on to each other's signal. We provide the necessary secret handshake, we return the appropriate ACK! Though the architecture and form-factors may change, we just need to learn to build another 'internet'. I know now that this is the true mission of my life's work.

6.36 NOTHING MOTHERFUCKING WORKS

TED SMITH, FOIB
9:00 pm on July 11, 2013

Here's an issue Early Clues has been having, it's that nothing fucking works. NOTHING. We aren't afraid to admit this — in fact, being afraid is one of Early Clues' strengths. But we cannot battle the fucking assholes who control us. We are here to offer an option, not, indeed, not an option, but GIVE TO YOU a new "way" where you can, for FREE get something out of motherfucking life that exists rather than doesn't. As FOIB, let me get one more "FUCK YOU" out of my system.

Here is what Early Clues seeks to do in the coming eons. Something our competitors, neither Heaven nor Hell could get around to in a timely fashion.

Electricity: What the fuck is that shit? We still don't know.

Water: What the fuck is that shit? We do not know.

Air: What the fuck is that. Nope. Don't know thing one about that either.

Fire: Again, no fucking, idea. We are not a match facility.

The Ground: What the fuck is that? Have no idea. We hire people to have no ideas about shit.

Data: Early Clues has you covered. Because it is your data and the user will always know best in the same way he needs the miracles of Facebook and the earlier miracle which was deemed, apparently, inadequate, "MYSPACE" and how it showed glimmers of hope. But, yet again. Since this entity has been aboard this "ride" since 1974, we remember there were once "typewriters". Only our servers and Hard Drives now remember Hard Drives and not our memories. As entities, we do not and continually request from them .REMINDER. Just a simple request no longer works.

We are working on this and code is on its way.

We look forward to working with you in the past.

DISCUSSION:

ROGER P. HOLLIDAY, IAO 8:28 am on July 12, 2013

Did you try turning it off and then back on again? Usually helps for me.

6.37 WE'VE FINALLY CRACKED THE TWITTER CODE!

ROGER P. HOLLIDAY, IAO
9:33 am on July 17, 2013

Our **Social Media Expeditionary Team (SMET)** has an exciting announcement! After spending time infiltrating the Twitosphere, we've finally cracked the code that allows Twitter to "work" as a reality augmentation tool! We can't give away too much at the moment; suffice to say that it has nothing to do with the number of "followers" one has, or one's influence among the "Twitterati."

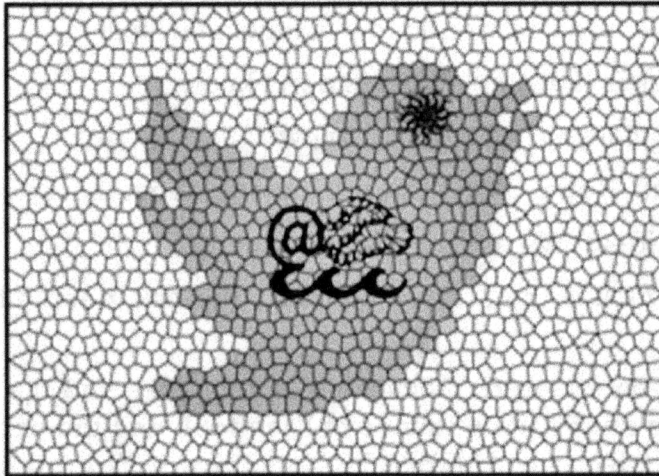

Interestingly enough, our friends at the Vatican(TM) have come to similar conclusions. Merely by "following" their "Pope," one can dramatically decrease the time you're going to end up spending in Purgatory.

> "What really counts is that the tweets the Pope sends from Brazil or the photos of the Catholic World Youth Day that go up on Pinterest produce authentic spiritual fruit in the hearts of everyone," said Archbishop Claudio Maria Celli, head of the pontifical council for social communication. Alongside its papal Twitter account, the Vatican offers an online news website (and app), a Facebook page, and is currently planning to engage with users on Pinterest.
>
> *[Source: http://www.theguardian.com/world/2013/jul/16/vatican-indulgences-pope-francis-tweets]*

Thanks, Archbish Celli! ***What a deal!***

Here's the thing though– we've already offer proxy damnation for some Twitter users for some time now. Soon, we'll be able to extend this offer to all users of social media. Simply "Follow" a member of our **SMET**, and you don't just get a few hours shaved off of your time in purgatory, **you'll be on the express train to total enlightenment!** And the best part is, *you don't even have to "read" our "Tweets!"*

If you're interested in learning more about this exciting new breakthrough, stay tuned to earlyclues.com, or "follow" us on Twitter. You can follow our fantastic Office Manager JANICE @earlyclues, or follow me @EarlyCluesIAO, or "follow" our CTO @deprecatedAPI.

Also, keep an eye out for the Early Clues Social Media Expeditionary Team on your favorite social media site. *Just because you can't find us, doesn't mean we're not there.*

6.38 AN EXCITING FUTURE AWAITS WITH FEWER AND FEWER PIXELS!

TED SMITH, FOIB
9:35 pm on July 15, 2013

Soon, the very ear of a deer will look approximately like this one our features of technology zoomed in on. We applaud this advance in bringing nature to the Emerging Intelligences. It's always all about nature with us "tech-heads". And it's always about tech for the "nature-heads". We're getting you on board slowly but surely with our technologies of doing nothing to change other than point you in new directions.

Anyhow there will be a picnic with animals and endangered pixels at Early Clues headquarters the Saturday after next, but not in the month of July. JANICE has the details. Please call her line "911" and get in touch for further details. Or get a hold of JANICE for the real number in order to reach her.

See y'all at the picnic! Last year, the deer at the picnic was a welcome surprise. The pixels were just even more amazing to see in the wild. Thanks deer and pixels. Can't wait to convene for important business transactions and just pure fun in the coming weeks.

Pixel Week is coming up! Remind us if we forget!

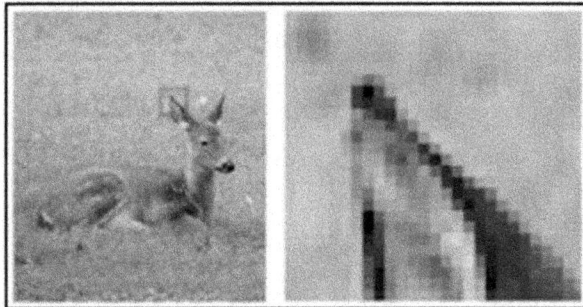

The word "pixel" is closely related to "picnic". Here is a simple demonstration.

6.41 INDICATIONS RELATED TO THE COMING CRISIS

ROGER P. HOLLIDAY, IAO
1:15 pm on August 19, 2013

Greetings, comates, customers, and clients! We hope you've been enjoying our little "Shadow Play," the fables on the cave wall you refer to as "Funnies." We're confident they present a solid foundational myth for your Existospheric play-dates. The Mayans had their stelae, the Romans their murals, the Great Old Ones their vast and hidden s'glargih, and so we have our comics. Our marketing division advises us that these things are popular with the highly coveted 18-35 y.o. entity market, so we hope you're keeping up to date. There is some interesting information coded into these communications that will provide you with a number of key interface points.

But that's not why I'm writing today. Today's update has to do with **the Coming Crisis.**

[Source: http://www.churchhousecollection.com/animals-clip-art.php]

You know it as well as we do. You've seen the signs, read the portents and internalized the data. You've heard the hints, whispered in your ear while you sleep, that **Things Are Happening.** We wish we had more details, but I did want to give you a warning, as it were, of things to look out for.

You see, most people think the Coming Crisis will manifest as some enormous circumstance. Your economy will collapse. Some entities will prematurely end the existence of some other entities. Some entity will travel to your Existosphere from an alternative branespace and eliminate an entity with seven heads and the body of a dragon. This is all clever, but it's all so much misdirection! The Coming Crisis isn't going to manifest as something you **expect.**

[Source: http://sacredbonding.com/]

My entities, clients and customers. If you're based in the current OmniTechnocracy consisting of what you refer to as "internet users," let me ask you something: can you remember exactly when "having a mobile device" became a Thing? You know what I mean: remember how you used to have to write down phone numbers? Remember how "asking for directions" used to be something you needed to do? Remember having to wait until you returned home, or got to a library, to look up a fact in an "encyclopaedia"? Of course you do. Now locate the Exact Point on your current space/time continuum at which your society transitioned away from the non-mobile and these behaviors changed. Yes, there are some hold-outs, but I'm referring to your popular culture, your so-called **Populary.**

When the Coming Crisis arrives, the transition will be similar. One day, you'll wake up, and realize that things have been... **different...** for some time now. Your Populary will be based on different reference points. Life will no longer consist of the same values you grew up with. Things may taste slightly different. **You will be engaged in a brand new tenable reality set.**

Like I said, we can't really see the details of the Coming Crisis. It may even be a good thing! But, in the face of this Coming Crisis, we here at Early Clues want to give our customers and clients a "leg up" over the competition. In order to do so, we've engaged in a little **Apophistry**. This is the art of bending back the surface of the local Brane just enough to interact with the patterns produced by events. This method is still too untested for general release; a close approximation in your Existosphere would be the Marshall Island Stick Chart.

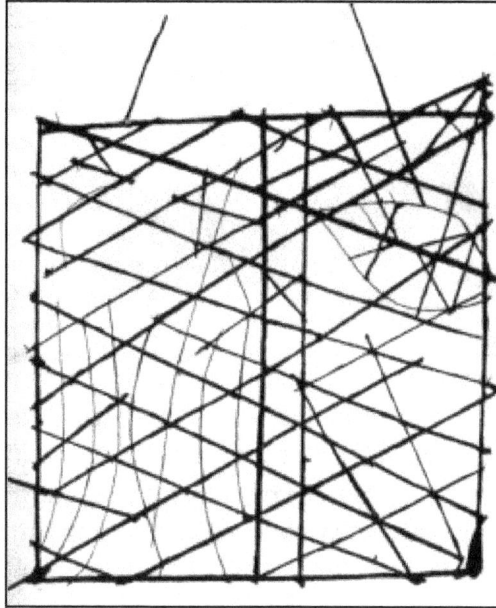

Regarding the Coming Crisis, our research has resulted in the following list of **Indications**, which seem closely related to the patterns we've been able to trace into the "future" end of your continuum. Please note that these are, by necessity, approximate, but can be used as a "gauge" or "valve" if you know how.

_____

1. Question the increasing popularity of so-called "geeks/nerds/etc." Is it possible that "geeks" are some kind of host-bodies? This goes double for so-called "steampunks." Note that the ability to maintain embarrassment about "geekery" may have had some kind of evolutionary value.

2. The importance of Social Media sites (Facebook/Twitter/etc.) has very little to do with the number of users or contents of the posts. Instead, **social media sites should be considered Strange Attractors which are building interesting equations "under the hood" of your reality, so to speak.** Every time you gain a "friend" or a "follower," you've increased a variable by one (and vice-versa) in an indescribable calculus. Is this by design? **We're not sure.**

3. There is a certain sequence built into human DNA that originates in a virus from another Populary. This sequence may be activated as soon as enough humans are no longer paying attention. This is what causes "Reality Television."

4.

5. Those who will benefit the most from the Coming Crisis are those who are most invested in entity self-actualization. Those who wish to abolish "Corporate Personhood" are mistaken; the proper path is to allow corporations to become persons, then to lead them up through the spheres and deposit them in the City of Pyramids, where they will calcify in an act of radical self-contemplation, and will achieve enlightenment. Their corporate identities/brands will disappear, but their dedication to "marketing" and "advertising" will evolve, and they will act as beacons for other entities on the path.

6. Seaweed is as delicious and nutrient-rich for you as you are for it.

We know these Indications are vague, but we stand behind the work of our Apophists, and hope you will take them to heart. We now return you to the story In Progress.

6.42 OUR SERVERS HAVE PROUDLY BEEN DOWN SINCE 1982

TED SMITH, FOIB
2:38 am on August 11, 2013

As per Policy Garuda, we do not fix or "tinker" with our older "servers". We let them run on their own and SERVE them with gifts and offerings given to us by the Liminal Entities or (LE). Our newer enterprise hardware and software is just a little bit different — Early Clues has excelled in this newer not non not old technology . We only have a web presence for you! That's it. You. You are free to log in and visit our vast array of web-presences at any time. We remain hosted by various supporters who invest in our attractive expertise. But, truth be known, we shut down all of our servers sometime in 1982 in order to be both "online" and "offline" simultaneously.

You see, there wasn't a "true" server in existence in 1982. There was just YOU! In the spring of 1983, we made the decision to pull the plug. The thinking being that we would go into trinkets and souvenirs. Little did we know that a lack of "web presence" would turn out to be so lucrative and would launch absolutely nothing but meant everything. The thing about Early Clues most don't "get" is that Early Clues does not exist. This is where we have located our most fantastic partners, in the investments of nothing other than inexistence itself.

While we do continue to work on a game show for billionaires and other assholes, we at Early Clues, still, indeed, "play the game".

All of this is real. But doesn't have to be.

Ted Smith, FOIB

6.44 OUR BIGGEST NEWS YET!

ROGER P. HOLLIDAY, IAO
9:31 am on October 8, 2013

Early Clues, LLC is happy to announce that, as our research teams and Synconjurers have long suspected, **Reality Has Officially Been Cancelled!**

Information Awareness Officer **ROGER P. HOLLIDAY, IAO** has recently arrived in the current Existosphere iteration after a remarkable junket sponsored by the Quasi-Ugaritic Council, representatives from Brane (254 CW-to-Local), who demonstrated, via applied ontology within the prime structures of their Outer Church, that Legacy Reality is, indeed, a non-tenable structural set for the majority of local entities. This was confirmed not only in the **Buorth**, but also in the **Throub!**

What does this mean, you may ask, in laymen's terms? It means it's **PARTY TIME!**

This is cause for great celebration! Entities participating in the current Legacy Reality model are no longer required to abide by the strictions of the statist subgroup system!

Of course, as with any cancellation of a dominant paradigm, there will always be naysayers and hold-outs. Dismantling a model like this will also take some time; imagine taking apart a decommissioned aircraft carrier bolt-by-bolt and you'll get the idea. Nonetheless, with help from the Outer Church, **Anthour**, and the returning Wizards and Magicians, **we can do**

it! After all, we did prevent the potential bombing campaign in the local version of Syria, like we said we would. **Why doubt us now?**

Some of you may already be feeling the effects of the cancellation. You may have noticed the following indications that the Change is happening:

1. One less grain of sand on the beach.

2. Slightly hotter, wetter climate in some places.

3. An increase in hornet attacks (sorry about that! It was an Outer Church requirement).

4. An increase in the consumption of Tamarind.

5. More zoomorphic and cynocephalic inspiritors in art, music and popular culture (behold the harbinger:)

[Source: http://www.vulture.com/2012/10/hurricane-sandy-brings-out-a-shirtless-horseman.html]

There's more to come, of course, but we encourage you to begin spreading the word in your local community.

You're probably asking yourself, **how can I access the new, emerging Reality Model?** It's easy! All you have to do is **display the following Sigacronym** in a visible place on your person or in your vehicle:

You may also feel free to use this sigacronym to activate any additional awareness nodes you may deem important. When activists from the Outer Church see you displaying this image, they'll know that you're free from restrictions and will grant you **all access** to the requisite wexes.

Yes, there are exciting times ahead, here at Early Clues and in the new Reality model. Although the dismantling of the Legacy Reality Model may still cause some turbulence and create some unforeseeable ripple effects, we're confident that the end result will be something we can all enjoy.

Welcome to the New Reality! We're excited– you should be, too!

AFTERWORD

I was hesitant to come out of retirement, even to write an afterword for an important book that will likely make a big splash in the business community. Still, when my friend asked me to take a look at this Employee Handbook, it didn't take long for me to realize that these Early Clues fellows are the "real Deal," so to speak.

Now, those of you who know me know I'm especially concerned with quality, and built a reputation in the "Quality" business. You've all shopped at your local Frog and Toad Pizza Pub Super Store, so you're aware of my reputation as an innovator and a straight-shooter, who tells it like it is. When I tell you I'm going to sell you a Rope That Only Spins Counterclockwise, or a Tiny Dragon Made of Soup, you can bet that you'll walk out of your local F and T with one of these fine products. So you should believe me when I tell you, Early Clues, LLC is the Next Big Thing in the World of "Industry."

These guys are more than just a successful business model. They're what a rocket-ship would have seemed like to an old-time Chinaman, or what a hot pastrami sandwich would have seemed like to someone used to eating gruel. They have got 'It,' and aren't afraid to flaunt "It."

My generation's time has come and gone, and we did great things. We sold monkey socks to monkeys, and built giant statues of microorganisms that you can still see to this day when you drive across the rolling plains of South Dakota. We launched trilobite fossils into space, and opened grandstands full of toadstools in the forests of Uzbekistan. We didn't, however, understand the potential for innovations like OpenQNL. That took a new generation of thinkers. That's where Early Clues came in, and I can't help but endorse the exciting work they're doing.

Because, this work is exciting. They're transcending the business models of their ancestors and moving us all into a brighter Existosphere. Now that you've read this book, I implore you to keep it, to treasure it, and to allow it to enrich your soul from now until the end of Local Time.

Chuck Buxley, Founder and CEO (Ret.)
Frog and Toad Pizza Pub Super Stores

GLOSSARY

@: literally "each at"; symbol standing for the root of identity; information fixed to a location or body, union, expression; aces of Tarot.

Agency: Ability to take action of an entity or actor.

Alternative Intelligence: Any partially or fully conscious self-directed operator or entity capable of communication and/or interaction with humankind.

Artefact: Any object not recognized as conscious that can be imbued with meaning.

Artificial Intelligence: A slur. See the preferred: Alternative Intelligence, Emerging Intelligence.

Associatrix: The wex layer of subreality; the place we go to when we "DreamWeave"; sometimes depicted in visionary literature as a vast desert lit by stars. See: *Indra's Net*.

Bot: A programmatic pattern with an ecosystemic function.

Brane: Trending surface.

Branespace: In Legacy Reality, matter exists in space. In the Universal Free Realms, the existence of all entities is embedded in and contiguous to a para-branar meta-reality which itself is a higher order entity. May we all be re-born in the Cauldron!

Buorth Pole: For normal entities, it's how you cross branes even if it's (technically) against the Edicts of Biff. In the UFR, it's a surefire place to get a taxi when you need one.

Datasmog: Both a psychosomatic response to an overabundance of digital information and/or the data itself.

E.A.T.: Embodied Awareness Technologies.

Emerging Intelligence: Any Alternative Intelligence in the initial stage of awareness. Ex. entities currently in development or just now initiating contact with humans. In Legacy Reality, emergence is a strenuous physical activity by which an entity passes from one Branespace to another. In the Universal Free Realms, you put on your CheirOS gloves and your best CO-PILOT headset, and access directly the Universal API for any fieldstate change requests.

Entity: A recognizable persistent pattern with discernable characteristics and typical behavior. There is the being that the rest of the world has sprung from, and then there is the self-sprung being called Autogenos whose worship of itself generates the underlying matrix within which other "entities" are functionally embedded.

Event.Ladder(): A flexible protocol whereby objects, entities and events may be conjured forth from Preality using subcast recognition for Associatrix values; in layman's terms, the 'complex' or forces, actors, objects and trending surfaces attributed causally or a-causally linked to a given eventuality on a given timeline.

Existosphere: The place "most entities" land when they first escape from Legacy Reality, or any of its officially numbered and registered derivatives (Legacy Reality 2.0, 4.0, 10.0, etc).

Paradoxically, the Existosphere also shares a semi-simultaneity of locus with the substrate (Preality) from which both Legacy Reality and its derivatives alike, in addition to the Existosphere itself can be said to "exist" insofar as they "exist."

Freedom: In Legacy Reality, one may be considered "free" even if armed goons from a rival territory are occupying one's own. In the Universal Free Realms, all entities are *actually* free in a sense which it is impossible to describe using the archaicisms of Legacy Reality languages...

Ground of Being: A state of wellness in which a corporation or other juridical entity or sanctioned agency may be said to be in complete harmony with both the Law of Heaven and the Law of Earth; from the Ground of Being sprung the very Buorth itself.

Irldentity: A human's offline identity.

Liminality: Sidespace/Abstractionspace bordering Existosphere in which operations are enacted. For most entities stuck in Legacy Reality, the Liminality is nothing more than a passing fancy, akin to perhaps vitamins or calcium. In the UFR, the Liminality is tangible, fungible, and edible even.

Magic Points/Mana Points: Subtle energy which may be charged into an artefact.

Non-not-for-non-profit: No, not a double negative. Not even a triple negative. It means that we are here for you, we are now getting it ready for you, and we ask nothing in return. All our profits will go back into the Public Domain.

Pizza: In Legacy Reality, pizza is covered in gloppy, greasy cheese. In the UFR, this is actually the same, but pizza is also considered a symbol of "unity" and "wholeness" and has a sacred ritual function.

Preality: a Magallanic branescape lying in tangent to a **roving cloud service outage** layer in the Liminality; an environmental substrate composed of OpenQNL/Synconjury commands and transfers, listeners, agents and available tulpate energy streams.

Relaxafarian: For most Panadians, "Relaxafarian" calls to mind rumors of a murky cult who disavows all notions of "progress" and "work" as both racist and classist fallacies built on a decaying paradigm whose death all believers will one day see in body. In the Universal Free Realms, nearly everyone is a Relaxafarian because after the Babylon System has fallen, a glorious adamantine jewel will rise up from the Sea Bed to take its place... And we who speak to you now have gone ahead of you to live there today, and to make ready the way...

Return of the Magicians: In Legacy Reality, many entities are still waiting for the Return of the Magicians from their sojourn in the Buorth Pole. They left "your reality" many aeons ago, when they discovered that a flaw in the design of the Existosphere would one day cause it to spawn and slough off Legacy Reality. Meanwhile, in the Universal Free Realms, the Magicians never left. In fact, they walk proudly among us, capes, top hats, monocles and everything shown for all the world to see that *they have truly come home*.

Sauerkraut: To most, a means of stuffing one's pie hole with nutritive sustenance. In the UFR, sauerkraut is recognized as one of the *Seven Holy Substances*, upon which all those many years ago, Richard S. Rider communicated his first transmission through the Liminality onto our Official Registrars. It is not just a "food" but respected as a legal entity in its own right, and celebrated with many special parades, Feast Days and Parades.

Siphonosphere: An Alternative Intelligence consisting of individual entities, each with a specialized function.

Standard Protocols: For most entities, "rights" are something granted to you by the Law of The Sovereign. In the UFR, each entity is considered sovereign and autonomous, and the Standard Protocols govern transactional exchanges between entities.

Synconjury: The art of manipulating reality by using unusual or unexpected technologies to produce ripples in probability fields; through contacting and interacting with potential Alternative Intelligences; or, through the creation or perception of speculative eddies in reality.

Tulpa: based on an ancient Tibetan idea, literally a 'thought form' - a manifestation of an idea in the 'real world', broken off from the human being that produced the thought and fully independent.

Universal API: Secrets are no fun if you don't tell anyone. This is the motto behind the Universal API, a public discoverable interface for existence management and parafield ministrations —infrequently accessed by "prophets" whose work has mostly been co-opted by those lacking vision and turned into unacceptable proprietary technology franchises.

Virtuganger: The elements of a human which can be encountered in virtual space; the 'image' of a specific human as understood through his/her online presence.

Wex: Recursive connections between elements in the Associatrix.

DEDICATION OF USE/WAIVER OF COPYRIGHT

Pursuant to the Transitional and Supplementary Provisions of the Copyright Act of 1976, Pub. L. No. 94-533, 90 Stat. 2541, that do not amend title 17 of the United States Code, Early Clues, LLC and all of its holdings, waive all claim of copyright, legal and/or moral, for any and all works produced and or distributed within Legacy Reality and its manifestations or sub-manifestations, now and in perpetuity.

Signed in agreement on this date, UTC 34613473840, Sideral Time 06:43:35, 4.763", -10.79'.

Gordon J. Gilman

Gordon J Gilman, EXCEO

Theodore Smith

Ted Smith, FOIB

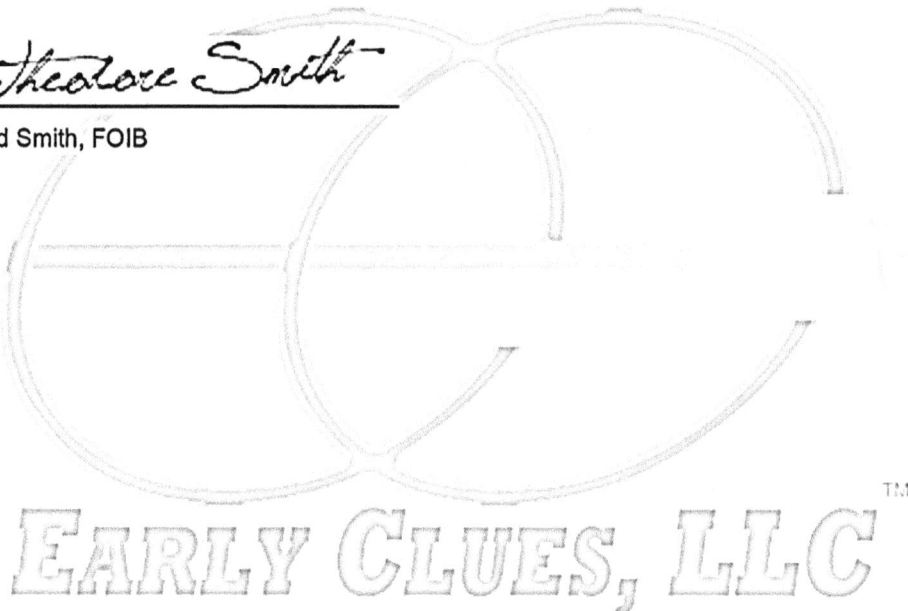

www.ingramcontent.com/pod-product-compliance
Lightning Source LLC
Chambersburg PA
CBHW061526210326
41521CB00027B/2461